Towards Green, Enhanced Photocatalysts for Hydrogen Evolution

Towards Green, Enhanced Photocatalysts for Hydrogen Evolution

Editors

Andrea Speltini
Antonella Profumo

MDPI • Basel • Beijing • Wuhan • Barcelona • Belgrade • Manchester • Tokyo • Cluj • Tianjin

Editors

Andrea Speltini
Department of Chemistry
University of Pavia
Pavia
Italy

Antonella Profumo
Department of Chemistry
University of Pavia
Pavia
Italy

Editorial Office
MDPI
St. Alban-Anlage 66
4052 Basel, Switzerland

This is a reprint of articles from the Special Issue published online in the open access journal *Catalysts* (ISSN 2073-4344) (available at: www.mdpi.com/journal/catalysts/special_issues/photocatalysts_hydro_evolution).

For citation purposes, cite each article independently as indicated on the article page online and as indicated below:

LastName, A.A.; LastName, B.B.; LastName, C.C. Article Title. *Journal Name* **Year**, *Volume Number*, Page Range.

ISBN 978-3-0365-1748-3 (Hbk)
ISBN 978-3-0365-1747-6 (PDF)

Contents

About the Editors vii

Preface to "Towards Green, Enhanced Photocatalysts for Hydrogen Evolution" ix

Antonella Profumo and Andrea Speltini
Towards Green, Enhanced Photocatalysts for Hydrogen Evolution
Reprinted from: *Catalysts* **2021**, *11*, 93, doi:10.3390/catal11010093 1

Panagiotis Marios Adamopoulos, Ioannis Papagiannis, Dimitrios Raptis and Panagiotis Lianos
Photoelectrocatalytic Hydrogen Production Using a TiO_2/WO_3 Bilayer Photocatalyst in the Presence of Ethanol as a Fuel
Reprinted from: *Catalysts* **2019**, *9*, 976, doi:10.3390/catal9120976 3

Lan Luo, Tingting Zhang, Xin Zhang, Rongping Yun, Yanjun Lin, Bing Zhang and Xu Xiang
Enhanced Hydrogen Production from Ethanol Photoreforming by Site-Specific Deposition of Au on Cu_2O/TiO_2 p-n Junction
Reprinted from: *Catalysts* **2020**, *10*, 539, doi:10.3390/catal10050539 15

Xiaobin Qiu, Lingfang Qiu, Mengfan Ma, Yingying Hou and Shuwang Duo
A 3D Hierarchical Pancake-Like Porous Carbon Nitride for Highly Enhanced Visible-Light Photocatalytic H_2 Evolution
Reprinted from: *Catalysts* **2020**, *10*, 77, doi:10.3390/catal10010077 33

Andrea Speltini, Lidia Romani, Daniele Dondi, Lorenzo Malavasi and Antonella Profumo
Carbon Nitride-Perovskite Composites: Evaluation and Optimization of Photocatalytic Hydrogen Evolution in Saccharides Aqueous Solution
Reprinted from: *Catalysts* **2020**, *10*, 1259, doi:10.3390/catal10111259 47

Zubia Saleem, Erum Pervaiz, M. Usman Yousaf and M. Bilal Khan Niazi
Two-Dimensional Materials and Composites as Potential Water Splitting Photocatalysts: A Review
Reprinted from: *Catalysts* **2020**, *10*, 464, doi:10.3390/catal10040464 59

About the Editors

Andrea Speltini

Andrea Speltini is Assistant Professor at the University of Pavia. After his Ph.D. research, which focused on the application of carbon nanotubes in separation science, in particular as solid-phase extraction sorbents and chromatographic stationary phases, as a post-doc researcher he has been involved in the development of analytical methods for the determination of contaminants and pharmaceuticals in complex matrices. A consistent part of the research has been conducted in the field of environmental chemistry, in particular to explore alternative photocatalytic systems for hydrogen gas evolution from water, in the presence of sacrificial (waste)biomasses. He is currently working on carbon materials for the sample treatment of environmental, food and biological matrices, and on the application of novel composite catalysts based on TiO_2, graphitic carbon nitride, carbon dots and perovskite for H_2 photogeneration under solar light.

Antonella Profumo

Antonella Profumo is Full Professor of Analytical Chemistry and Director of the Department of Chemistry of the University of Pavia. Her main research activities, in collaboration with her research group of young researchers, focus on the development and application of selected analytical procedures for preconcentration on carbon-based solid phases in different matrices (environmental, food, and biological samples); the use and application of analytical procedures in the field of forensic pedology; and the employment of novel catalysts for Hydrogen generation by sacrificial water splitting in the presence of renewable biomasses. She runs scientific collaborations with public and private foundations that refer to the Department for analytical guidance, in particular in the sector of natural spring and drinking waters, and in the environmental field as far as concerns emissions of inorganic and organic pollutants from industrial plants and the recovery of material from waste in a circular economy perspective.

Preface to "Towards Green, Enhanced Photocatalysts for Hydrogen Evolution"

Nowadays, more and more people are becoming aware that environmental protection is a priority. The constant growth of energy demand combined with the adverse effects on the global environment caused by use of fossil fuels is increasingly requiring clean and renewable energy sources. The most desired goal is producing energy in new ways, possibly exploiting green sources, in the framework of a *circular economy*. To this aim, *photocatalysis* represents a smart, energy-saving, and very promising technology that relies on the use of an abundant energy source as solar light. In the recent years, *hydrogen photocatalytic production* by water splitting and aqueous-phase reforming of different organics has been under deep study by many research groups because H_2 gas is a highly desired energy vector that produces water by combustion, thus perfectly matching the requirements of green chemistry. Although most of the available data derive from lab-scale studies, it seems believable that, in the near future, H_2 could be obtained on a large scale by making use of inexhaustible, green sources such wastewater and biomass derivatives with the aid of another inexhaustible resource as sunlight. However, this necessarily requires that *efficient, recyclable* and *safe* photocatalysts are designed. The authors of this work are involved in environmental studies including the development of alternative H_2-producing photocatalytic systems, so the aim of this book is to gather some of the recent studies in the context of *hydrogen photocatalytic production*. This topical cluster could be interesting for researchers working in the fields of materials science, applied photocatalysis and environmental sciences.

Andrea Speltini, Antonella Profumo
Editors

catalysts

MDPI

Editorial

Towards Green, Enhanced Photocatalysts for Hydrogen Evolution

Antonella Profumo [1,*] and Andrea Speltini [2,*]

1 Department of Chemistry, University of Pavia, Via Taramelli 12, 27100 Pavia, Italy
2 Department of Drug Sciences, University of Pavia, Via Taramelli 12, 27100 Pavia, Italy
* Correspondence: antonella.profumo@unipv.it (A.P.); andrea.speltini@unipv.it (A.S.)

Citation: Profumo, A.; Speltini, A. Towards Green, Enhanced Photocatalysts for Hydrogen Evolution. *Catalysts* **2021**, *11*, 93. https://doi.org/10.3390/catal11010093

Received: 8 January 2021
Accepted: 8 January 2021
Published: 12 January 2021

The constant growth of energy demand joined with the adverse effects on the global environment induced by use of fossil fuels is increasingly requiring new routes to obtain clean and renewable energy sources. Hydrogen photocatalytic production by water splitting or aqueous-phase reforming of different organics has been under deep investigation for some years; H_2 gas is the most desirable energy vector because the only product by combustion is water. H_2 could be obtained at the large scale in the near future by exploiting inexhaustible, green sources such as sunlight, wastewater and biomass derivatives, if efficient, recyclable and safe photocatalysts will be designed.

Besides a review article, this Special Issue gathers research on the preparation, characterization and application of new organic/inorganic composites endowed with photo(electro)-catalytic properties. These materials were tested under either UV–visible or, even more conveniently, under visible light for H_2 evolution in "sacrificial water splitting" or "photoreforming" systems.

A new 3D porous carbon nitride catalyst has been proposed by Qiu et al. [1]. The material was synthesized by a simple bottom-up procedure entailing self-assembly of a melamine–cyanuric acid complex followed by ethanol and glycerol intercalation prior to calcination. This route has the merit to yield a 3D hierarchical pancake-like highly porous carbon nitride with enhanced light-harvesting capacity, expanded band gap, prolonged charge carriers' lifetimes, and higher surface area and reduction ability towards hydrogen ions to produce gas-phase H_2, compared to the bulk material. Under visible-light radiation and platinum as a co-catalyst, the hydrogen evolution rate (HER) from triethanolamine aqueous solution was 430 μmol $g_{cat}{}^{-1}$ h^{-1}, 9-fold larger than that afforded by non-porous carbon nitride. The semiconductor showed photochemical stability. Indeed, it was successfully reused for three additional photoreactions. The behavior of the novel catalyst was studied in water containing a fine chemical, as a proof-of-concept sacrificial agent. Three other research studies focused on H_2 evolution from water in the presence of more sustainable electron donors, such as saccharides and alcohols [2–4], testing new catalysts as well.

Carbon nitride–perovskite composites, which presently denote a cutting-edge research field, were investigated as new photoactive micro-sized materials for H_2 evolution from glucose aqueous solution as a representative sacrificial biomass [2]. In particular, the synergism between the newer lead-free perovskite and carbon nitride, due to improved charge carrier separation derived from the positive band-alignment between the two semiconductors, has been exploited to set up a sacrificial water splitting system working under simulated solar light. The H_2 production was optimized by a design of experiments, achieving an HER higher than 900 μmol $g_{cat}{}^{-1}$ h^{-1}, 12-fold larger compared to pure carbon nitride. The best-performing composite also provided H_2 with no metal co-catalyst, and in the presence of untreated starch, selected as an abundant and low-cost biopolymer, and therefore exploiting the sacrificial role of a raw polysaccharide. Despite the lower surface area, the perovskite–carbon nitride composite results were attractive compared to nanometric P25 TiO_2, relative to H_2 evolution.

A representative biomass-derived substrate such as ethanol has been selected in the research of Luo et al. [3] to demonstrate the feasibility of a more sustainable method of obtaining H_2. Indeed, the contemporary production of bioethanol has reached several tens of million tons per year, and the possibility of ethanol photoreforming under mild conditions is of great relevance. This paper by Luo shows that coupling selective decoration with Au nanoparticles of TiO_2 nanorods and the TiO_2/Cu_2O p–n junction produces H_2 along with acetaldehyde. The latter is stoichiometrically formed because the C–C cleavage of ethanol does not occur, resulting in no release of greenhouse gases such as carbon dioxide. Under simulated solar light, the new photocatalytic system Au@TiO_2/Cu_2O affords an HER higher than 8500 μmol $g_{cat}{}^{-1}$ h^{-1} over the composite Au@Cu_2O/TiO_2 and Au@TiO_2, and it maintains unchanged performance for at least five consecutive catalytic runs.

In the report by Adamopolous and co-workers [4], a nano-sized TiO_2/WO_3 bilayer catalyst was employed as a photoanode in a photoelectrochemical cell to produce H_2 from aqueous ethanol solution. The system proposed, which involved photoelectrocatalytic alcohol reforming, takes advantage of the high oxidative power and visible light absorption of WO_3, used in the photoanode, stabilization of charge carriers by electron-transfer from TiO_2 to WO_3, and passivation of tungstate surface states, which reduces the number of charge recombination sites involved by the titania layer. The latter constituent also scatters back to the bottom WO_3 layer part of the incident light resulting in a higher photocurrent production, proportional to the applied voltage, thus in greater H_2 formation at the cathode of the cell.

This Special Issue also includes a review article focused on the application of 2D materials and composites as potential photocatalysts for water splitting [5]. With more than 200 references covering the last two decades, but with particular attention to the papers published in recent years, after providing the reader with some introductory sections summarizing the fundamentals of water splitting photocatalysis, the application of various materials is presented and discussed through comprehensive tables reporting key information about each photocatalytic system (e.g., catalyst band gap, light source, type and amount of co-catalyst, sacrificial agent used, and HER). The paper covers selected studies on graphitic carbon nitride and graphene-based photocatalysts, metal phosphides, metal organic frameworks and derivatives, together with those on the more recent phosphorene. The review emphasizes the progress in modern nanomaterial applications, for instance by metal nanoparticles doping, surface functionalization, synthesis-controlled morphology, which are essential to achieve the most desired properties, i.e., low charge recombination, high light harvesting capability, good electron conductivity, fast kinetics, and large surface area. From the survey by Saleem and co-authors, it emerged that the use of 2D materials, their combinations and derivatives, are now at the basis of further advancements in photocatalytic water splitting.

We take the chance to thank the authors and their co-authors for the contributed publications to this Special Issue.

Funding: This research received no external funding.

Conflicts of Interest: The authors declare no conflict of interest.

References

1. Qiu, X.; Qiu, L.; Ma, M.; Hou, Y.; Duo, S. A 3D hierarchical pancake-like porous carbon nitride for highly enhanced visible-light photocatalytic H_2 evolution. *Catalysts* **2020**, *10*, 77. [CrossRef]
2. Speltini, A.; Romani, L.; Dondi, D.; Malavasi, L.; Profumo, A. Carbon nitride-perovskite composites: Evaluation and optimization of photocatalytic hydrogen evolution in saccharides aqueous solution. *Catalysts* **2020**, *10*, 1259. [CrossRef]
3. Luo, L.; Zhang, T.; Zhang, X.; Yun, R.; Lin, Y.; Zhang, B. Enhanced hydrogen production from ethanol photoreforming by site-specific deposition of Au on Cu_2O/TiO_2 p-n junction. *Catalysts* **2020**, *10*, 539. [CrossRef]
4. Adamopoulos, P.A.; Papagiannis, I.; Raptis, D.; Lianos, P. Photoelectrocatalytic hydrogen production using a TiO_2/WO_3 bilayer photocatalyst in the presence of ethanol as a fuel. *Catalysts* **2019**, *9*, 976. [CrossRef]
5. Saleem, S.; Pervaiz, E.; Yousaf, M.U.; Niazi, M.B.K. Two-dimensional materials and composites as potential water splitting photocatalysts: A review. *Catalysts* **2020**, *10*, 464. [CrossRef]

catalysts

MDPI

Article

Photoelectrocatalytic Hydrogen Production Using a TiO_2/WO_3 Bilayer Photocatalyst in the Presence of Ethanol as a Fuel

Panagiotis Marios Adamopoulos, Ioannis Papagiannis, Dimitrios Raptis and Panagiotis Lianos *

Department of Chemical Engineering, University of Patras, 26500 Patras, Greece; cmng3238@upnet.gr (P.M.A.); ion.papg@gmail.com (I.P.); dgraptis86@yahoo.gr (D.R.)
* Correspondence: lianos@upatras.gr; Tel.: +30-2610-997513

Received: 13 September 2019; Accepted: 19 November 2019; Published: 21 November 2019

Abstract: Photoelectrocatalytic hydrogen production was studied by using a photoelectrochemical cell where the photoanode was made by depositing on FTO electrodes either a nanoparticulate WO_3 film alone or a bilayer film made of nanoparticulate WO_3 at the bottom covered with a nanoparticulate TiO_2 film on the top. Both the electric current and the hydrogen produced by the photoelectrocatalysis cell substantially increased by adding the top titania layer. The presence of this layer did not affect the current-voltage characteristics of the cell (besides the increase of the current density). This was an indication that the flow of electrons in the combined semiconductor photoanode was through the WO_3 layer. The increase of the current was mainly attributed to the passivation of the surface recombination sites on WO_3 contributing to the limitation of charge recombination mechanisms. In addition, the top titania layer may have contributed to photon absorption by back scattering of light and thus by enhancement of light absorption by WO_3. Relatively high charge densities were recorded, owing both to the improvement of the photoanode by the combined photocatalyst and to the presence of ethanol as the sacrificial agent (fuel), which affected the recorded current by "current doubling" phenomena. Hydrogen was produced under electric bias using a simple cathode electrode made of carbon paper carrying carbon black as the electrocatalyst. This electrode gave a Faradaic efficiency of 58% for hydrogen production.

Keywords: WO_3; TiO_2; hydrogen production; photoelectrocatalysis

1. Introduction

Photoelectrocatalytic hydrogen production is one of the most popular research subjects because it promises an effective route for converting solar energy and storing it as chemical energy in the form of hydrogen. The popularity of hydrogen stems from the fact that it has the highest gravimetric heat of combustion (~286 kJ/mole), while its combustion leads to the production of water. Hydrogen is mainly produced by reforming of fossil fuels. It can also be produced by electrolysis using, for example, renewable electricity. Theoretically, 1.23 V are necessary to split water by electrolysis; however, in reality, much higher voltages and expensive electrocatalysts are necessary. In this sense, water splitting by photoelectrocatalysis [1] is a very promising approach since it necessitates much lower electric biases. However, oxidation of water is a four-electron process:

$$2H_2O \rightarrow O_2 + 4H^+ + 4e^- \quad (1)$$

Therefore, it is necessary to simultaneously extract four units of charge in order to oxidize water and produce molecular oxygen. This is not easy and necessitates oxygen evolution co-catalysts. On the contrary, it is easier to oxidize an organic substance. For example, in the case of ethanol, oxidation leads to the formation of acetaldehyde, which is a two electron process, therefore easier to realize, either directly [2–4]:

$$C_2H_6O \rightarrow CH_3CHO + 2H^+ + 2e^- \quad (2)$$

or by the intermediate of hydroxyl radicals:

$$OH^- \rightarrow {}^{\bullet}OH + e^- \text{ and } C_2H_6O + 2{}^{\bullet}OH \rightarrow CH_3CHO + 2H_2O \quad (3)$$

For most organic substances, oxidation and subsequent mineralization proceed by steps, as in the above case of ethanol, which usually are two electron processes, therefore easier to oxidize than water. In terms of oxidation potential, oxidation of ethanol to acetaldehyde corresponds to −0.21 V [5], i.e., it is much lower than the above mentioned potential of 1.23 V necessary to oxidize water. Molecular hydrogen production by reduction of protons or water is also a two electron process [3,4]:

$$2H^+ + 2e^- \rightarrow H_2 \text{ (acidic environment)} \quad (4)$$

or

$$2H_2O + 2e^- \rightarrow 2OH^- + H_2 \text{ (alkaline environment)} \quad (5)$$

Therefore, photoelectrocatalytic hydrogen production by using an organic substance as a fuel is even more of a promising approach, which has been adopted in the present work.

The standard configuration of a photoelectrocatalytic cell for hydrogen production involves a photoanode electrode carrying a semiconductor photocatalyst and a cathode electrode where reductive reactions are carried out, producing hydrogen. Both electrodes are immersed in an electrolyte, which increases the internal cell conductivity and provides the ionic species necessary to allow device functioning in a cyclic manner [3,4]. Figure 1 shows a simplified design of a photoelectrocatalysis cell. When photons are absorbed by the photocatalyst, electron-hole (e^--h^+) pairs are generated. Holes are consumed by oxidation reactions, such as (2) or (3), while electrons move through an external circuit and are consumed by reduction reactions at the cathode. The overall result of oxidation and reduction can be represented, for example, by combining (2) or (3) with (4) or (5), respectively [2]:

$$C_2H_6O \rightarrow CH_3CHO + H_2 \quad (6)$$

which describes the consumption of ethanol, from now on referred to as the "fuel", to produce hydrogen by means of a photoelectrocatalytic process. Acetaldehyde may be further oxidized until complete mineralization and further hydrogen production. The complete mineralization of ethanol may be expressed by the following overall reaction [3,4]:

$$C_2H_6O + 3H_2O \rightarrow 2CO_2 + 6H_2 \quad (7)$$

which is the same as ethanol reforming [6] and can be rightfully called "photoelectrocatalytic alcohol reforming".

Photoelectrocatalysis necessitates then the presence of a photocatalyst. There exists a rich literature on the choice of photocatalysts [4,7,8]. Researchers searching for a new photocatalyst have developed resourceful arguments in their favor; however, it is a matter of fact that very few among them simultaneously possess more than one quality that distinguishes each one of them from the others. Nanostructured titania is the best such example, but WO_3 is also a distinguished choice as well [9–11]. Titania is an n-type, low cost, non-toxic semiconductor, easy to process, and having a strong adhesive quality on various types of electrodes. It possesses a strong oxidative potential (about 3 V in the case of anatase) and a conduction band conveniently located with respect to the hydrogen production potential (about −0.2V vs. NHE) [12]. Titania also possesses relatively large charge carrier mobility. For example, its hole diffusion length is of the order of 10^4 nm, much larger than that of WO_3 (~150 nm) and hematite (2–4 nm) [13]. These properties would make titania an ideal photocatalyst; however, it still suffers from substantial electron-hole recombination, while its light absorption range is limited only to the UV. Nanoparticulate tungsten oxide is also a very popular choice, and thanks to its substantial range of visible light absorption, it has been studied for several decades [11,14–17] as an alternative to the UV absorbing titania. Indeed, WO_3 has a bandgap ranging between 2.5 and 2.8 eV, and it may thus absorb light up to 500 nm, which accounts for 12% of the solar radiation on the surface of the Earth [17]. WO_3 is an n-type indirect semiconductor. It is easy to synthesize and deposit on electrodes; it has a moderate hole-diffusion length (~150 nm [17], as already said); it is resistant against photocorrosion; and it is stable at relatively low pH values. For this reason, WO_3 has been studied as a photoanode material for photoelectrochemical water splitting applications [11,14–20]. Its valence band is located approximately at +2.8 V vs. NHE [12]; therefore, it also possesses high oxidative power, and it is well placed for water and organics' oxidation. Its conduction band is located at positive potentials (approximately +0.2 to +0.3V vs. NHE, cf. [12]); therefore, it is located substantially lower than that of titania. If the two semiconductors are then brought into contact, it is expected that photogenerated electrons may be transferred from TiO_2 to WO_3, thus achieving charge separation and limiting e^--h^+ recombination. However, it is understood that since the photon absorption spectral range for titania is limited, its role as an electron source for WO_3 is of limited importance. There is another role that titania may play, which is of utmost importance, and this is the passivation of surface states of WO_3 that a top cover of titania can offer. This role will be investigated in the present case.

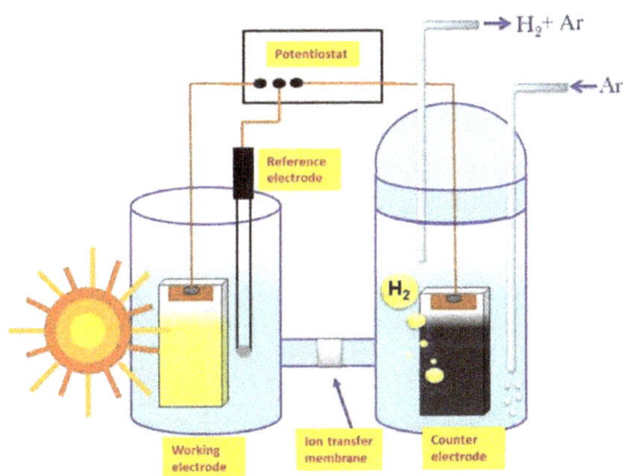

Figure 1. Schematic representation of the reactor employed in the present work, showing the working electrode (photoanode), counter electrode (cathode electrode carrying an electrocatalyst), reference electrode, and connections with a potentiostat, which provides the necessary bias. Hydrogen production is monitored in a de-aerated environment obtained by Ar flow.

In addition to the surface states' passivation effect, there is one more reason to enhance WO_3's performance. As will be discussed below, contrary to the transparent WO_3 layer, the top titania layer is a light scattering film, which scatters back to WO_3 at least part of the incident light. This situation supports conditions for enhancement of light absorption by the WO_3 layer, and this, of course, results in higher photocurrent production.

Unbiased flow of electrons from the photoanode to the cathode electrode would necessitate a substantial potential difference between the two electrodes. The potential of the photoanode is roughly defined by the conduction band of the photocatalyst, while that of the cathode electrode by the redox potential of the reduction reaction. In the case of hydrogen production, the potential in a strongly acidic environment is around 0 V vs. NHE. The conduction band of titania lies at about −0.2 V vs. NHE, as already said. This difference of 0.2 V is too small, and it is eliminated by losses. It is even worse in the case of WO_3, which has a conduction band slightly more positive than the hydrogen production potential. Therefore, no unbiased flow of electrons is expected for a WO_3 photoanode and is hardly expected in the case of a TiO_2 photoanode. In other words, when the above two photocatalysts are used, production of hydrogen may be obtained only by applying a bias. This is the reason that in the present work, hydrogen was produced under electric bias, as schematically illustrated in Figure 1.

By taking into account the above knowledge, in the present work, we are studying photoanodes made by combining nanoparticulate WO_3 with nanoparticulate TiO_2, in a bilayer structure comprising a bottom film made of WO_3 and a top film made of TiO_2, in order to investigate the merits of such a combination. Measurements have focused on photoelectrocatalytic hydrogen production by using ethanol as the model fuel [19] with the understanding that other organic materials, either biomass byproducts or water soluble pollutants, may apply as well [20].

2. Results and Discussion

In the present work, as already said, the photoanodes were made either by deposition of a WO_3 nanoparticulate film alone on an FTO electrode or by adding a layer of nanoparticulate titania on the top. As seen in Figure 2A, WO_3 nanoparticles were polydisperse with the size ranging between 20 and 50 nm. The BET specific surface area was 26 m^2g^{-1}, and X-ray diffractograms (not shown) revealed the formation of monoclinic WO_3 crystallites of size around 25 nm as calculated by using Scherrer's formula. The quantity of WO_3 in the film was approximately 4.5 mg cm^{-2}. When the titania layer was added on the top, a complete coverage of the lower WO_3 film was obtained by exposing only the titania film. The SEM image of the film in Figure 2B then showed a typical titania nanostructure. The quantity of titania was approximately 4 mg cm^{-2}. The latter maintained the characteristics previously described in other publications [21,22]. The thickness of the combined semiconductor film was approximately determined by its scanning electron microscope profile, and it was below 1 μm.

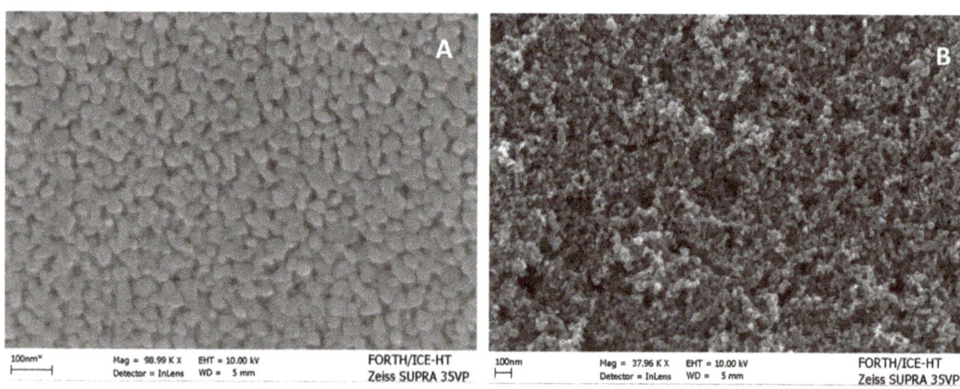

Figure 2. FE-SEM images of a WO_3/FTO (**A**) and a TiO_2/WO_3/FTO (**B**) film.

The spectroscopic characteristics of the WO_3 film before and after titania deposition are shown in Figure 3A. WO_3 film alone absorbed light up to about 480 nm, i.e., within a spectral range typical of this semiconductor. When titania was added on the top, the reflectance-absorption spectrum presented a different aspect depending on which side of the film was exposed to light. When the light passed through the transparent FTO electrode (cf. Figure 3B), the spectrum (Curve 2) showed similar characteristics as those of WO_3 alone (Curve 1). Since the photoanode operates by this mode (Figure 3B), it was seen that the spectroscopic characteristics of WO_3 prevailed. When absorption was recorded by front reflection, i.e., by means of the top titania layer, the corresponding curve (Curve 3) had a typical aspect of a nanocrystalline titania film. This is an additional indication of complete coverage of the WO_3 layer by (the light scattering) titania.

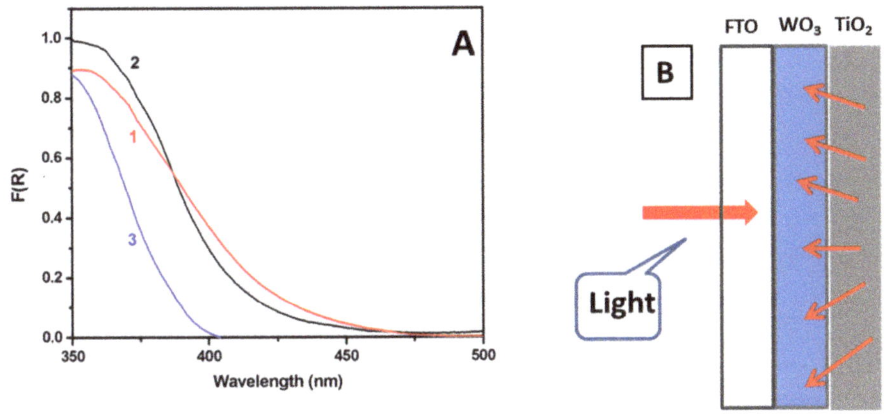

Figure 3. (**A**) Reflection-absorption spectra of the WO_3 (1) and the TiO_2/WO_3 (2,3) photoanode. Curve 3 was obtained by front side reflection and Curve 2 by reflection through the supporting FTO glass. Curve 3 is typical of the nanoparticulate titania absorption spectrum. (**B**) Graphical representation of the direction of photons in the presently used setup.

The presence of TiO_2 on the top of WO_3 was expected to encourage electron-hole separation, thanks to the difference between the levels of the conduction band of these semiconductors, which may reach more than 0.4 V, with the titania conduction band being the more negative of the two (cf. the band levels in [23]). However, as already said, the role of titania as a sensitizer of WO_3 was expected to be limited since its photon absorption range was limited. On the contrary, deposition of titania on the top of the WO_3 film may result in the passivation of surface defects of the latter, and this would limit charge recombination sites. Passivation of surface defects has been observed and recorded in TiO_2/WO_3 and other semiconductor combinations [24–30]. In addition, titania films are more stable than WO_3 films; therefore, the top titania layer provided protection against aggressive electrolytes. Finally, the top titania layer provided a light scattering film, in contrast to the transparent WO_3 film, which may send light back to WO_3, thus enhancing its capacity for photon harvesting. This is especially convenient in the present case where light was absorbed through the supporting FTO layer. This configuration is illustrated by the design of Figure 3B. Light scattering by the titania layer was due to the agglomeration of titania nanoparticles into large clusters (cf. Figure 2B).

The beneficial effects that the bilayer photocatalyst film had on the activity of the photoelectrocatalytic cell were immediately seen by recording current-voltage curves, as seen in Figure 4. More than a 30% increase of current was recorded when the titania layer was added on the top, compared with a photoanode made of WO_3 alone. The curves of Figure 4 were recorded under the same conditions, differing only by the composition of the photocatalyst film. The curves were recorded in light chopping mode demonstrating that the voltage range of pure photocurrent extended between about 0.2 and 1.8 volts vs. Ag/AgCl. The small anodic peak below this range was due to capacitance current deriving from the adsorption of cations from the electrolyte into the pores of the mesoporous film [21]. The increase of the dark current above 1.8 V vs. Ag/AgCl was due to electrolysis. The electrolyte was 0.5 M $NaClO_4$ with 5% v/v ethanol added in the anode compartment. In accordance with previous publications [19,20], in the absence of ethanol (not shown; cf. the data in [19]), lower currents were obtained in both photoanode cases. This was expected as previously explained [19,20]. Ethanol acted as a hole scavenger, which further limited charge recombination, thus resulting in higher currents. Moreover, transient species formed by ethanol oxidation induced current increase by the "current doubling" phenomenon [31]. Indeed, the maximum current density expected for a photoanode with light absorption threshold at 480 nm (cf. Figure 3A) was no more than 5 mA cm^{-2} [4]. However, the current density presently increased under bias well beyond this limit, and this was rationalized only by the current doubling effect. The addition then of the titania layer and the presence of the fuel (ethanol) resulted in the production of substantial photocurrent.

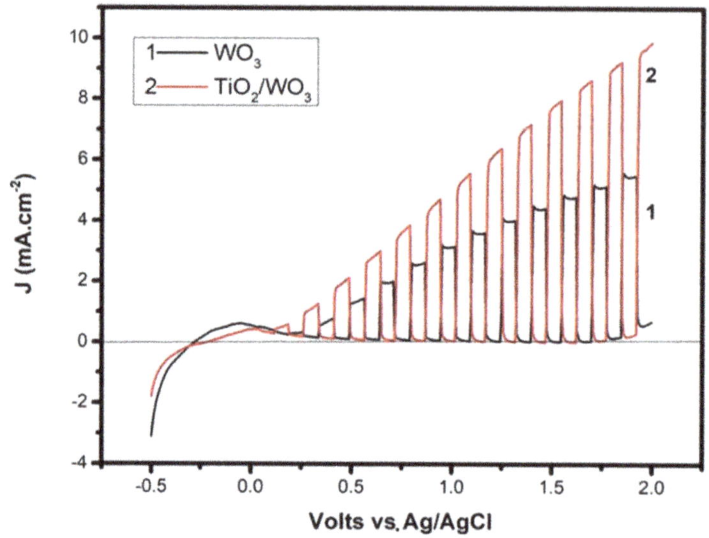

Figure 4. Current density-voltage curves for a photoelectrocatalytic cell made of two different photoanodes. Recording was done in light chopping mode demonstrating the voltage range of the pure photocurrent.

The increase of the current when titania was added on the top of WO_3 was not due to titania acting as the photocatalyst by itself. When a photoanode was made by using titania alone, by taking care to deposit the same quantity of photocatalyst (4 mg cm^{-2}), under the same conditions of the cell configuration and the electrolyte and fuel concentration, the behavior of the current-voltage curve presented a completely different aspect, as seen in Figure 5. The maximum current density was only 0.42 mA cm^{-2}, while in the case of Figure 4, the current increase was more than 3 mA cm^{-2}. Most importantly, open circuit voltage (the point where the current became zero) was <-1 V vs. Ag/AgCl, while in Figure 4, the open circuit voltage was approximately located at 0.0 V vs. Ag/AgCl. The difference in the onset potential between a WO_3 and a TiO_2 photoanode roughly reflected the difference in the electrochemical potential of each semiconductor, and it was systematically observed. The fact that the addition of titania on the top of WO_3 preserved the characteristics of the latter simply means that the bottom lying WO_3 layer defined the level at which electrons left the anode electrode. It was furthermore observed that in the case of titania alone, the photocurrent soon reached saturation by the voltage increase, while in the case of WO_3 and TiO_2/WO_3, the current continuously increased with bias voltage. In other words, the addition of the titania layer on the top of WO_3 resulted only in a current increase without affecting the original appearance of the pure WO_3 current-voltage curve. Obviously, photogenerated electrons in the titania layer were transferred to the electrode only by means of the underlying WO_3 layer, and titania did not act as a photocatalyst by itself. Because of the continuous increase of current with increasing voltage and since hydrogen production was proportional to the current flowing through the external circuit of the cell, it was expected that the hydrogen production rate should strongly depend on the applied electric bias.

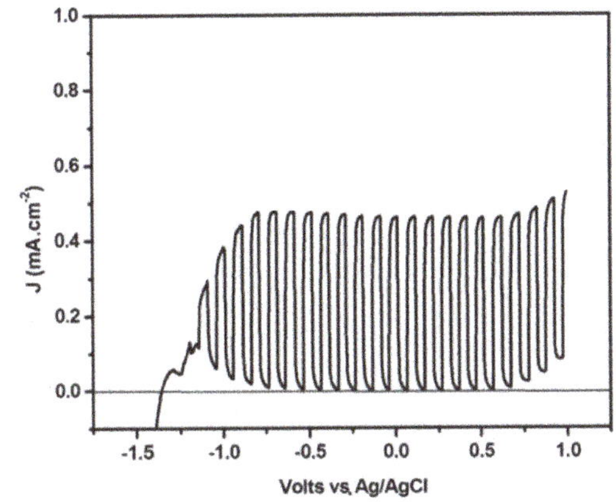

Figure 5. Current density-voltage curves for a photoelectrocatalytic cell made of a (pure) titania photoanode. Recording was done in light chopping mode demonstrating the voltage range of the pure photocurrent. The anodic rise at negative voltages is due to the capacitance current deriving from the adsorption of Na cations from the electrolyte solution [21].

At this point, a question arises about the fate of the photogenerated holes. An approximate positioning of band levels for TiO_2 and WO_3 nanoparticulate semiconductors and of the oxidation potentials of the three possible hole scavengers in the present system is presented by the diagram in Figure 6 (cf. also the band levels in [23]). Titania nanoparticles used for the present application were a mixture of anatase and rutile. For this reason, the valence band level could not be clear cut and was expected to lie between 3.0 and 2.8 V vs. NHE. A similar situation existed for WO_3 due to the fact that its bandgap may range between 2.5 and 2.8 eV, as already said. This created an uncertainty concerning the direction of the hole transfer. In Figure 6, we favored the transfer of holes to titania, which, being on the top, facilitated the interaction with hole scavengers in the solution. In view of structural imperfections in both semiconductors, it was hard to make a strong assertion on this matter. In any case, the level of 3.0 V vs. NHE was positive enough to produce $^{\bullet}OH$ radicals. However, vast losses, again due to structural imperfections, may make transfer of holes to OH^- ions less probable than the most obvious hole scavenging by ethanol or water. Oxidation of ethanol, as already said, is easier than water therefore, ethanol remained the most probable hole consumer.

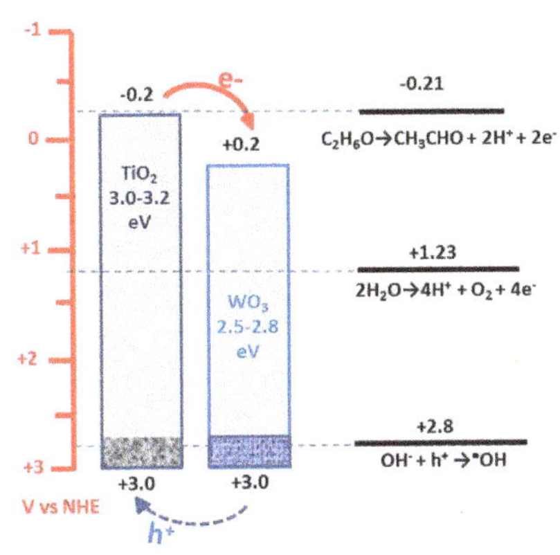

Figure 6. Approximate band levels for nanoparticulate TiO_2 and WO_3 and oxidation potentials of the possible hole scavengers in the present system.

The hydrogen production rate by employing the two types of photoanode was indeed monitored, and the results are presented in Figure 7. In Figure 7, we tried to demonstrate the strong variation of the hydrogen production rate as a function of the applied bias. No hydrogen production was observed without a forward bias. Three different electric biases were applied, chosen within the above photocurrent range. Thus, each curve was divided into three parts, one for each applied bias, as marked on the graph. Each part demonstrated a period of a rising rate followed by a period of a practically constant rate. A fast increase of the hydrogen production rate took place after changing the applied bias, and it corresponded to the filling of tubes and detection device with the additional hydrogen produced. The variation of the hydrogen production rate with applied bias was a clear demonstration that hydrogen produced by reduction reactions at the cathode depended on the current flowing through the external circuit of the cell. Indeed, by comparing Curve #2 of Figure 4 with Curve #2 of Figure 7, it is found that the hydrogen produced was roughly proportional to the current. In addition, the combined TiO_2/WO_3 photoanode produced a higher quantity of hydrogen than the WO_3 photoanode, in accordance with the higher current flow. The maximum hydrogen production rate recorded under the present conditions, with 1.6 V bias vs. Ag/AgCl, was 4.4 μmol.min^{-1}. The corresponding current was 24.3 mA. The Faradaic efficiency for hydrogen production under the present conditions could be calculated by the following equivalent [32]: 1 μmole/min H_2 corresponds to (10^{-6} moles) × (6.022 × 10^{23} molecules/mole) × (2 electrons/H_2 molecule) × (1.6 × 10^{-19} Coulomb/electron)/60 s, which is equal to 3.21 mA. Consequently, 4.4 μmole.min^{-1} correspond to 14.1 mA. Since the average real current was 24.3 mA, then the Faradaic efficiency was 58% (14.1/24.3 = 0.58). The Faradaic efficiency depended on the device configuration and reactor parameters, as well as the efficiency of the cathode electrode and the quality of the electrocatalyst. The presently obtained value was relatively low, owed mainly to the use of an inexpensive electrocatalyst (carbon paper loaded with carbon black), but also to the fact that the reactor parameters were not optimized. Comparable Faradaic efficiencies were also obtained in a previous work [32]. It must be stressed at this point that the quality of the photoanode and the strength of the applied bias mainly dictated the flowing current, while the functionality of the cathode electrode mainly affected the Faradaic efficiency. It relied then mainly on the counter electrode to improve Faradaic efficiency, and this will be a matter of our future projects.

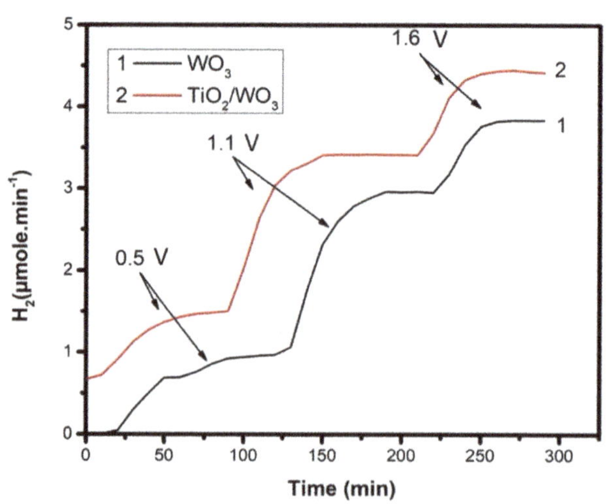

Figure 7. Variation of the photoelectrocatalytic hydrogen production rate by applying various electric biases. Biases are marked on the graph, and they are expressed in volts vs. Ag/AgCl electrode. The two curves correspond to the two photoanodes.

The data of Figure 7 corresponded to relatively short-time measurements. When the experiment was run for a longer period of time, a slow current drop was observed, which was accompanied by an analogous drop of the hydrogen production rate (cf. the data in our previous work [32]). The proportionality of current with the hydrogen production rate preserved the value of the Faradaic efficiency, in accordance with the above discussion. It was then again concluded that the photoanode and its immediate environment, as well as the applied bias dictated the current production, while the capacities of the counter electrode dictated the Faradaic efficiency for hydrogen production.

3. Materials and Methods

3.1. Materials

Reagents were obtained from Sigma Aldrich and were used as received. Millipore water was used in all experiments. SnO_2:F transparent conductive electrodes (FTO, resistance 8 ohm/square) were purchased from Pilkington (Pilkington, Toledo, OH, USA) and carbon paper from SGL Technologies GmbH (Meitingen, Germany, thickness 190 μm), and carbon black was a donation from Cabot Corporation (Vulcan XC72, Billerica, MA, USA).

3.2. Construction of WO_3 Photoanodes

A quantity of 0.4 g of high purity tungsten powder (99.99%) with average particle sizes up to 10 microns reacted with 3 mL of 30% aqueous hydrogen peroxide [33] under sonication for 2–3 h until a transparent colorless solution was obtained. The excess of H_2O_2 was catalytically decomposed using a Pt foil [34,35] by keeping the mixture at 10 °C with a thermostated water bath. Then, the solution was mixed with 3 mL of ethanol and 0.3 g of the surfactant Triton X-100, and the mixture was deposited on previously cleaned FTO slides cut into the appropriate dimensions. FTO slides were cleaned first with soap and then by sonication in acetone, ethanol, and water. Deposition was made by spin-coating at 2000 rpm for 20 s. After depositing 3 subsequent layers, the obtained film was calcined for 10 min at 500 °C (heating rate of 20 °C/min). The procedure was repeated at least six times. This rather tedious procedure was necessary to make a compact film without voids. The active area of the film was 1 cm × 1 cm in the case when it was used for IV measurements. Larger films of 3.5 cm × 5 cm were employed for hydrogen production.

3.3. Preparation of the TiO_2 Paste and Deposition on the WO_3 Film

Deposition of titania on the top of the WO_3 film was achieved by using a paste made of commercial titania nanoparticles. The paste was prepared by modifying a previously published procedure [36]. Three grams of Degussa P25 powder were placed in a small mortar with 0.5 mL of acetic acid and carefully ground for 3 min. Then, 2.5 mL of triple-distilled water were added in 5 equal doses while grinding the mixture. Grinding continued by then gradually adding 10 times ethanol by 1 mL and subsequently 6 times by 1.25 mL. At the end of this procedure, a big amount of ethanol, almost 30 mL, was added, which helped to detach titania from the walls of the mortar. The mixture was then sonicated several times, for 2 s each time, until a homogeneous dispersion was obtained. Consequently, we added 10 g of 2-(4-methyl-1-cyclohex-3-enyl)propan-2-ol (terpineol) and repeated ultrasonication again until a homogeneous dispersion was obtained. Finally, we added an amount of ethyl cellulose (5% w/w) dissolved in EtOH (10% w/v) and again repeated ultrasonication to obtain a homogeneous mixture. The whole procedure lasted 50–60 min, and when it ended, the solution was ready for use. This paste can be condensed in a rotary evaporator; however, in the present work, it was used as is, in order to apply it by spin-coating. Indeed, titania was applied on either a WO_3 covered electrode or a plain FTO glass by spin-coating at 3000 rpm for 30 s.

3.4. Construction of the Counter Electrode

The counter electrodes that were used in the present work were made of carbon cloth on which a carbon black paste was deposited by the following procedure: 0.3 g of carbon black was mixed with 8 mL of distilled water by vigorous mixing in a mixer (more than 4000 r.p.m.) until it became a viscous paste. This paste was further mixed with 0.1 mL polytetrafluorethylene (Teflon 60% wt. dispersion in water) and then applied on a carbon paper cut into the appropriate dimensions. The paste was first dried at 80 °C, and then, it was sintered at 340 °C. The procedure was repeated so as to deposit 0.5 mg of carbon black per cm^2. The active area of the electrode was 17.5 cm^2 (5 cm × 3.5 cm).

3.5. Apparatus and Measurements

The apparatus used for photoelectrochemical measurements and hydrogen production is schematically presented by Figure 1. It was made of two Pyrex cylinders separated by an ion transfer membrane (ROBU, Germany, porosity SGQ 5, diameter 25 mm, thickness 2 mm). One cylinder accommodated the photoanode and the reference electrode (Ag/AgCl) and the other the counter electrode. The counter electrode compartment was sealed using appropriate fittings, which allowed de-aeration of the solution by Ar flow, collection of the produced H_2, and counter electrode electric connection. The photoanode compartment was exposed to the ambient environment. The active area of the counter electrode was 3.5 × 5 cm^2, while the area of the photoanode was 1 × 1 cm^2 in the case of current-voltage measurements, but increased to 3.5 × 5 cm^2 in the case of hydrogen production. Each compartment was filled with 200 mL of 0.5 M $NaClO_4$. However, in the anode compartment, we also added 5% v/v ethanol. Illumination was made in all cases using a Xe lamp providing an intensity of 100 $mW{\cdot}cm^{-2}$ at the position of the photoanode. Hydrogen was detected online by using Ar as the carrier gas and an SRI 8610C gas chromatograph. Calibration of the chromatograph signal was accomplished by comparison with a standard of 0.25% H_2 in Ar. Current-voltage curves and application of electric bias were accomplished with an Autolab potentiostat PGSTAT128N (Metrohm AG, Herisau, Switzerland)

4. Conclusions

This work demonstrated the improvement of the photoelectrocatalytic behavior and the increase of the hydrogen production rate by adding a titania film on the top of a WO_3 photoanode. The electrochemical characteristics of the photoanode were not modified by TiO_2 film addition, but the obtained photocurrent largely increased, while the quantity of hydrogen proportionally increased as well. This effect was mainly ascribed to the passivation of WO_3 surface states, thus decreasing the number of charge recombination sites. This result is in accordance with the observations described in [24]. Increased photocurrent (and hydrogen) production may additionally be ascribed to the TiO_2 light scattering effect.

Author Contributions: P.M.A.: investigation. I.P.: investigation. D.R.: investigation. P.L.: conceptualization.

Funding: This research received no external funding.

Conflicts of Interest: The authors declare no conflict of interest.

References

1. Pan, L.; Vlachopoulos, N.; Hagfeldt, A. Directly Photoexcited Oxides for Photoelectrochemical Water Splitting. *ChemSusChem* **2019**, *12*, 4337–4352. [CrossRef]
2. Sakata, T.; Kawai, T. Heterogeneous photocatalytic production of hydrogen and methane from ethanol and water. *Chem. Phys. Lett.* **1981**, *80*, 341–344. [CrossRef]
3. Lianos, P. Production of electricity and hydrogen by photocatalytic degradation of organic wastes in a photoelectrochemical cell the concept of the Photofuelcell: A review of a re-emerging research field. *J. Hazard. Mater.* **2011**, *185*, 575–590. [CrossRef]

4. Lianos, P. Review of recent trends in a photoelectrocatalytic conversion of solar energy to electricity and hydrogen. *Appl. Catal. B. Environ.* **2017**, *210*, 235–254. [CrossRef]
5. Sfaelou, S.; Lianos, P. Photoactivated Fuel Cells (PhotoFuelCells). An alternative source of renewable energy with environmental benefits. *AIMS Mater. Sci.* **2016**, *3*, 270–288. [CrossRef]
6. Kondarides, D.I.; Daskalaki, V.M.; Patsoura, A.; Verykios, X.E. Hydrogen production by photo-induced reforming of biomass components and derivatives at ambient conditions. *Catal. Lett.* **2008**, *122*, 26–32. [CrossRef]
7. Yao, T.; An, X.; Han, H.; Chen, J.Q.; Li, C. Photoelectrocatalytic Materials for Solar Water Splitting. *Adv. Energy Mater.* **2018**, *8*, 1800210. [CrossRef]
8. Li, Z.; Luo, W.; Zhang, M.; Feng, J.; Zou, Z. Photoelectrochemical Cells for Solar Hydrogen Production: Current State of Promising Photoelectrodes, Methods to Improve Their Properties, and Outlook. *Energy Environ. Sci.* **2013**, *6*, 347–370. [CrossRef]
9. Shen, S.; Chen, J.; Wang, M.; Sheng, X.; Chen, X.; Feng, X.; Mao, S.S. Titanium dioxide nanostructures for photoelectrochemical applications. *Prog. Mater. Sci.* **2018**, *98*, 299–385. [CrossRef]
10. Jian, J.; Jiang, G.; van de Krol, R.; Wei, B.; Wang, H. Recent advances in rational engineering of multinary semiconductors for photoelectrochemical hydrogen generation. *Nano Energy* **2018**, *51*, 457–480. [CrossRef]
11. Kalanur, S.S.; Duy, L.T.; Seo, H. Recent Progress in Photoelectrochemical Water Splitting Activity of WO_3 Photoanodes. *Top. Catal.* **2018**, *61*, 1043–1076. [CrossRef]
12. Gratzel, M. Photoelectrochemical cells. *Nature* **2001**, *414*, 338–344. [CrossRef] [PubMed]
13. Liu, X.; Wang, F.; Wang, Q. Nanostructure-based WO_3 photoanodes for photoelectrochemical water splitting. *Phys. Chem. Chem. Phys.* **2012**, *14*, 7894–7911. [CrossRef] [PubMed]
14. Hodes, G.; Cahen, D.; Manassen, J. Tungsten Trioxide as a Photoanode for a Photoelectrochemical Cell (PEC). *Nature* **1976**, *260*, 312–313. [CrossRef]
15. Ashokkumar, M.; Maruthamuthu, P. Photocatalytic Hydrogen Production with Semiconductor Particulate Systems: An Effort to Enhance the Efficiency. *Int. J. Hydrogen Energy* **1991**, *16*, 591–595. [CrossRef]
16. Alexander, B.D.; Kulesza, P.J.; Rutkowska, I.; Solarska, R.; Augustynski, J. Metal Oxide Photoanodes for Solar Hydrogen Production. *J. Mater. Chem.* **2008**, *18*, 2298–2303. [CrossRef]
17. Gan, J.; Lu, X.; Tong, Y. towards Highly Efficient Photoanodes: Boosting Sunlight-Driven Semiconductor Nanomaterials for Water Oxidation. *Nanoscale* **2014**, *6*, 7142–7164. [CrossRef]
18. Wang, S.C.; Tang, F.Q.; Wang, L.Z. Visible Light Responsive Metal Oxide Photoanodes for Photoelectrochemical Water Splitting: A Comprehensive Review on Rational Materials Design. *J. Inorg. Mater.* **2018**, *33*, 173–197.
19. Sfaelou, S.; Pop, L.C.; Monfort, O.; Dracopoulos, V.; Lianos, P. Mesoporous WO_3 photoanodes for hydrogen production by water splitting and PhotoFuelCell operation. *Int. J. Hydrogen Energy* **2016**, *40*, 5902–5907. [CrossRef]
20. Raptis, D.; Dracopoulos, V.; Lianos, P. Renewable energy production by photoelectrochemical oxidation of organic wastes using WO_3 photoanodes. *J. Hazard. Mater.* **2017**, *333*, 259–264. [CrossRef]
21. Pop, L.C.; Sfaelou, S.; Lianos, P. Cation adsorption by mesoporous titania photoanodes and its effect on the current-voltage characteristics of photoelectrochemical cells. *Electrochim. Acta* **2015**, *156*, 223–227. [CrossRef]
22. Sfaelou, S.; Raptis, D.; Dracopoulos, V.; Lianos, P. BiOI solar cells. *RSC Adv.* **2015**, *5*, 95813–95816. [CrossRef]
23. Choi, T.; Kim, T.-S.; Kim, J.H. Transparent nitrogen doped TiO_2/WO_3 composite films for self-cleaning glass applications with improved photodegradation activity. *Adv. Powder Technol.* **2016**, *27*, 347–353. [CrossRef]
24. Yang, M.; He, H.; Zhang, H.; Zhong, X.; Dong, F.; Ke, G.; Chen, Y.; Du, J.; Zhou, Y. Enhanced photoelectrochemical water oxidation on WO_3 nanoflake films by coupling with amorphous TiO_2. *Electrochim. Acta* **2018**, *283*, 871–881. [CrossRef]
25. Huang, W.; Wang, J.; Bian, L.; Zhao, C.; Liu, D.; Guo, C.; Yang, B.; Cao, W. Oxygen vacancy induces self-doping effect and metalloid LSPR in non-stoichiometric tungsten suboxide synergistically contributing to the enhanced photoelectrocatalytic performance of WO_{3-x}/TiO_{2-x} heterojunction. *Phys. Chem. Chem. Phys.* **2018**, *20*, 17268–17278. [CrossRef]
26. Zeng, Q.; Bai, J.; Li, J.; Xia, L.; Zhou, B.; Sun, Y. Highly-stable and efficient photocatalytic fuel cell based on an epitaxial $TiO_2/WO_3/W$ nanothorn photoanode and enhanced radical reactions for simultaneous electricity production and wastewater treatment. *Appl. Energy* **2018**, *220*, 127–137. [CrossRef]

27. Guaraldo, T.T.; Goncales, V.R.; Silva, B.F.; de Torresi, S.I.C.; Zanoni, M.V.B. Hydrogen production and simultaneous photoelectrocatalytic pollutant oxidation using a TiO_2/WO_3 nanostructured photoanode under visible light irradiation. *J. Electroanal. Chem.* **2016**, *765*, 188–196. [CrossRef]
28. Georgieva, J.; Valova, E.; Armyanov, S.; Philippidis, N.; Poulios, I.; Sotiropoulos, S. Bi-component semiconductor oxide photoanodes for the photoelectrocatalytic oxidation of organic solutes and vapours: A short review with emphasis to TiO_2-WO_3 photoanodes. *J. Hazard. Mater.* **2012**, *211*, 30–46. [CrossRef]
29. Georgieva, J.; Armyanov, S.; Poulios, I.; Sotiropoulos, S. An all-solid photoelectrochemical cell for the photooxidation of organic vapours under ultraviolet and visible light illumination. *Electrochem. Commun.* **2019**, *11*, 1643–1646.
30. Patel, P.P.; Ghadge, S.D.; Hanumantha, P.J.; Datta, M.K.; Gattu, B.; Shanthi, P.M.; Kumpta, P.N. Active and robust novel bilayer photoanode architectures for hydrogen generation via direct non-electric bias induced photo-electrochemical water splitting. *Int. J. Hydrogen Energy* **2018**, *43*, 13158–13176. [CrossRef]
31. Kalamaras, E.; Lianos, P. Current Doubling effect revisited: Current multiplication in a PhotoFuelCell. *J. Electroanal. Chem.* **2015**, *751*, 37–42. [CrossRef]
32. Doukas, E.; Balta, P.; Raptis, D.; Avgouropoulos, G.; Lianos, P. A Realistic Approach for Photoelectrochemical Hydrogen Production. *Materials* **2018**, *11*, 1269. [CrossRef] [PubMed]
33. Murau, P.C. Dissolution of Tungsten by Hydrogen Peroxide. *Anal. Chem.* **1961**, *33*, 1125–1126.
34. Orel, B.; Krasovec, U.O.; Groselj, N.; Kosec, M.; Drazi, G.; Reisfeld, R. Gasochromic Behavior of Sol-Gel Derived Pd Doped Peroxopolytungstic Acid (W-PTA) Nano-Composite Films. *J. Sol-Gel Sci. Technol.* **1999**, *14*, 291–308. [CrossRef]
35. Kudo, T. A New Heteropolyacid with Carbon as a Heteroatom in a Keggin-like Structure. *Nature* **1984**, *312*, 537–538. [CrossRef]
36. Ito, S.; Chen, P.; Comte, P.; Nazeeruddin, M.K.; Liska, P.; Pechy, P.; Gratzel, M. Fabrication of screen-printing pastes from TiO_2 powders for dye-sensitised solar cells. *Prog. Photovolt. Res. Appl.* **2007**, *15*, 603–612. [CrossRef]

catalysts

MDPI

Article

Enhanced Hydrogen Production from Ethanol Photoreforming by Site-Specific Deposition of Au on Cu_2O/TiO_2 p-n Junction

Lan Luo [1,†], Tingting Zhang [1,†], Xin Zhang [1], Rongping Yun [1], Yanjun Lin [1], Bing Zhang [2] and Xu Xiang [1,*]

1 State Key Laboratory of Chemical Resource Engineering, Beijing University of Chemical Technology, 15 Beisanhuan Donglu, Beijing 100029, China; luolanbuct@163.com (L.L.); ztt0729123@163.com (T.Z.); xiangxubuct@163.com (X.Z.); yuanrp@mail.buct.edu.cn (R.Y.); linyj@mail.buct.edu.cn (Y.L.)

2 School of Chemical Engineering, Zhengzhou University, 100 Science Avenue, Zhengzhou 450001, China; zhangb@zzu.edu.cn

* Correspondence: xiangxu@mail.buct.edu.cn

† These authors contribute equally.

Received: 15 April 2020; Accepted: 8 May 2020; Published: 13 May 2020

Abstract: Hydrogen production by photoreforming of biomass-derived ethanol is a renewable way of obtaining clean fuel. We developed a site-specific deposition strategy to construct supported Au catalysts by rationally constructing Ti^{3+} defects inTiO_2 nanorods and Cu_2O-TiO_2 p-n junction across the interface of two components. The Au nanoparticles (~2.5 nm) were selectively anchored onto either TiO_2 nanorods (Au@TiO_2/Cu_2O) or Cu_2O nanocubes (Au@Cu_2O/TiO_2) or both TiO_2 and Cu_2O (Au@TiO_2/Cu_2O@Au) with the same Au loading. The electronic structure of supported Au species was changed by forming Au@TiO_2 interface due to the adjacent Ti^{3+} defects and the associated oxygen vacancies while unchanged in Au@Cu_2O/TiO_2 catalyst. The p-n junction of TiO_2/Cu_2O promoted charge separation and transfer across the junction. During ethanol photoreforming, Au@TiO_2/Cu_2O catalyst possessing both the Au@TiO_2 interface and the p-n junction showed the highest H_2 production rate of 8548 μmol g_{cat}^{-1} h^{-1} under simulated solar light, apparently superior to both Au@TiO_2 and Au@Cu_2O/TiO_2 catalyst. The acetaldehyde was produced in liquid phase at an almost stoichiometric rate, and C−C cleavage of ethanol molecules to form CH_4 or CO_2 was greatly inhibited. Extensive spectroscopic results support the claim that Au adjacent to surface Ti^{3+} defects could be active sites for H_2 production and p-n junction of TiO_2/Cu_2O facilitates photo-generated charge transfer and further dehydrogenation of ethanol to acetaldehyde during the photoreforming.

Keywords: ethanol reforming; hydrogen production; Au catalyst; Ti^{3+} defect; site-specific deposition

1. Introduction

Hydrogen is extensively used in various industrial processes, e.g., in the petrochemical industry, metallurgy, fine chemical engineering, etc. [1]. Hydrogen, as a clean and renewable fuel, has aroused tremendous attention from both academic and industrial perspectives in the past decades because its energy-extraction process produces only water as a byproduct and emits no greenhouse gases, e.g., CO_2 or any pollutants [2,3]. At present, industrial production of hydrogen depends predominantly on steam reforming of CH_4, an energy-consuming process accompanied by CO_2 emissions. Therefore, it is highly desirable to develop efficient, economical and energy-neutral processes for sustainable H_2 production.

Ethanol could be produced in a sustainable way from huge-amount, low-grade biomass, e.g., lignocelluloses and agriculture waste besides from the conventional petrochemical route [4].

In addition, over several tens of million tons biomass-derived ethanol (bio-ethanol) is produced per year. Therefore, ethanol reforming provides an alternative and promising means of H_2 production from abundant and low-value biomass resources [5,6]. Photoreforming could be one of the most promising potential reforming ways because H_2 production is activated on catalysts by sunlight and is achieved conveniently at room temperature and ambient pressure [7–10]. Another advantage of ethanol photoreforming is that hydrogen in the gas phase and acetaldehyde in the liquid phase can be formed at a stoichiometric rate without CO_2 emission. Acetaldehyde can be used directly in many situations, or can be further converted into other chemicals. Exploiting high-efficiency catalysts is of great importance for achieving high activity in ethanol photoreforming under mild conditions.

Metal-supported TiO_2 photocatalysts have been extensively studied in alcohol photoreforming. Au-supported TiO_2 catalysts have been tuned with respect to metal particle size, annealing conditions in different gas atmospheres, and the phase structure of TiO_2 (anatase and rutile) [11]. The AuPd alloy was further studied for adjusting the electronic structure of metal component, resulting in enhanced activity in this reaction [12]. The films of metal/TiO_2 supported on glass were tested under UV light irradiation of ethanol/water mixtures, revealing that Pt was slightly more active than Au under those conditions [13]. It is recognized that the longitudinal SPR mode of Au is the main channel for transferring the hot electrons from Au toTiO_2 [14]. The selective deposition of TiO_2 or other electron acceptors at the tip of Au nanorods (NRs) or at the edge of Au nanodisks (NDs) results in highly active plasmonic photocatalysts [15]. Yang and co-workers studied the gold nanodisks and TiO_2 nanophases sandwich between zeolite nanosheets for hydrogen production by plasmonic photocatalytic reforming of methanol [16]. It has been reported that anatase TiO_2 is more active than the rutile one, and the higher surface availability of the former for Au is beneficial to H_2 production from ethanol under UV light irradiation. However, the significant amounts of other gaseous products (mostly CO, CO_2 and CH_4) were also released, probably by further photo-induced decomposition of ethanol [17].

It is highly desirable to explore alternative strategies of catalyst design for enhancing hydrogen production from bio-ethanol photoreforming and simultaneously inhibiting C–C cleavage of ethanol. Herein, we propose a strategy for constructing both p-n junctions between two kinds of semiconducting oxides and Au@TiO_2 interface by site-specific Au deposition for enhancing hydrogen production from ethanol reforming. The results indicate that the interaction of Au and the adjacent Ti^{3+} defects on Au@TiO_2/Cu_2O catalyst affects the electronic structure of Au. In addition, the p-n junction between Cu_2O and TiO_2 facilitates the charge separation and transfer across the interface. The simultaneous construction of the p-n junction and the metal/oxide interface on the supported catalyst results in a record-level H_2 production rate of 8548 μmol ${g_{cat}}^{-1}$ h^{-1} under simulated solar light and excellent recycling stability. Accompanied by hydrogen production, acetaldehyde at a stoichiometric rate is solely produced in the liquid phase, indicative of efficient inhibition of C–C cleavage during ethanol reforming.

2. Results and Discussion

The procedure for catalyst preparation is shown in Scheme 1. The experimental details are stated in Section 3 Materials and Methods.

High-resolution transmission-electron microscope (HRTEM) photographs are shown in Figure 1. Cu_2O nanocubes with an average size of ~40 nm were grown on TiO_2 nanorods with a lateral size of 80~120 nm. The close contact between them can be clearly observed and the lattice d-spacing of 0.243 nm and 0.31 nm is assigned to the plane (103) of anatase phase TiO_2 [18] and (110) of cubic phase Cu_2O [19], respectively (Figure 1a). The junction can be formed between two oxides having intimate contact [20]. The Au nanoparticles are site-specific deposited on either TiO_2 or Cu_2O or both, showing the same average size of 2.5 nm and high dispersion (Figure 1b–d). The Au loading is determined to be 0.85 wt% in the three catalysts Au@Cu_2O/TiO_2, Au@TiO_2/Cu_2O andAu@TiO_2/Cu_2O@Au by ICP analyses. The lattice distance of 0.235 nm corresponds to the plane (111) of cubic phase Au [21]. The Au nanoparticles display a well-defined spherical morphology and high crystallinity (Figure 1b–d).

The size and loading of Au nanoparticles change a little, whatever Au is deposited on any support. EDS elemental mappings show the homogeneous distribution of Ti and O in the nanorods and Cu and O in the nanocubes (Figure 1)

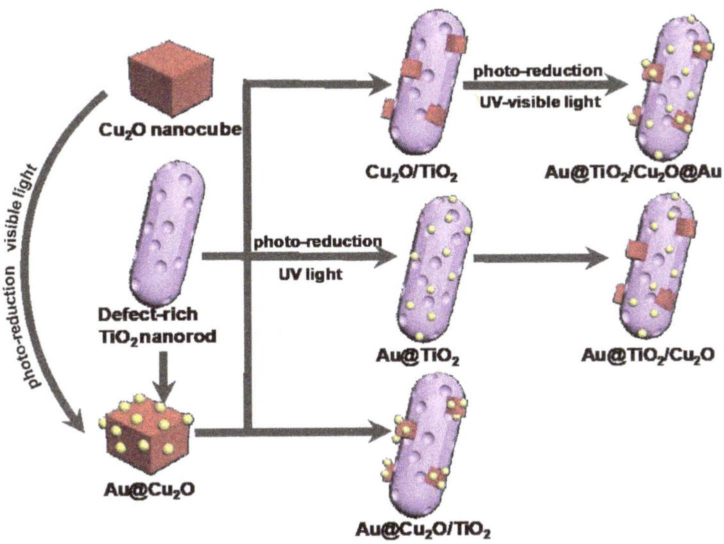

Scheme 1. The procedure for catalyst preparation by site-specific deposition of Au on the support.

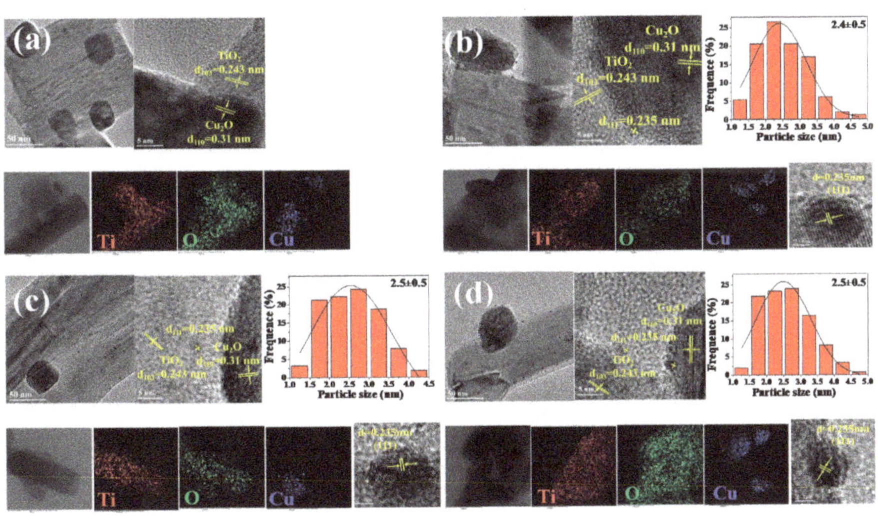

Figure 1. HRTEM images and EDS mapping of (**a**) Cu_2O/TiO_2 (**b**) $Au@Cu_2O/TiO_2$, (**c**) $Au@TiO_2/Cu_2O$, (**d**) $Au@TiO_2/Cu_2O@Au$, with histogram of Au NPs size distribution of each sample (100 particles are counted).

X-ray diffraction patterns of the samples indicate that TiO_2 exist in the form of anatase phase (JCPDS No. 21-1272). The second phase in the samples with p-n junction can be indexed to cubic Cu_2O (Figure S1). The diffraction related to Au phase is not observed in the Au-supported samples because of the low loading of Au (~0.85 wt%) and/or high dispersion of small-sized Au particles (~2.5 nm) on the surface of the support. X-ray photoelectron spectra (XPS) were used to analyze the chemical valence and electronic structure of each element. Two peaks were fitted at binding energy (B.E.) of 457.9 and 458.3 eV in the Ti $2p_{3/2}$ region of Cu_2O/TiO_2 and Au-supported catalysts, corresponding to Ti^{3+} and Ti^{4+} species, respectively (Figure 2A) [22–24]. The B.E. values hardly shift among these samples. However, the peak area ratio of Ti^{3+} to Ti^{4+} is lower when Au is deposited on TiO_2 i.e., $Au@TiO_2/Cu_2O$ and $Au@TiO_2/Cu_2O@Au$ (0.96 and 1.03), compared to that of Au-free TiO_2/Cu_2O and Au deposition on Cu_2O i.e., $Au@Cu_2O/TiO_2$ (1.17 and 1.11) (Table 1). This indicates that the site-specific deposition of Au differs at defect-rich TiO_2 and Cu_2O, leading to a variation to Ti^{3+} content when Au is deposited to TiO_2. In our previous studies, the ratio of Ti^{3+}:Ti^{4+} increased as the amount of $NaBH_4$ increased. When the ratio reaches around 1, even when increasing the amount of $NaBH_4$, the ratio will not increase any more, which suggests that it is approaching saturation of Ti^{3+}:Ti^{4+} ratio [25].

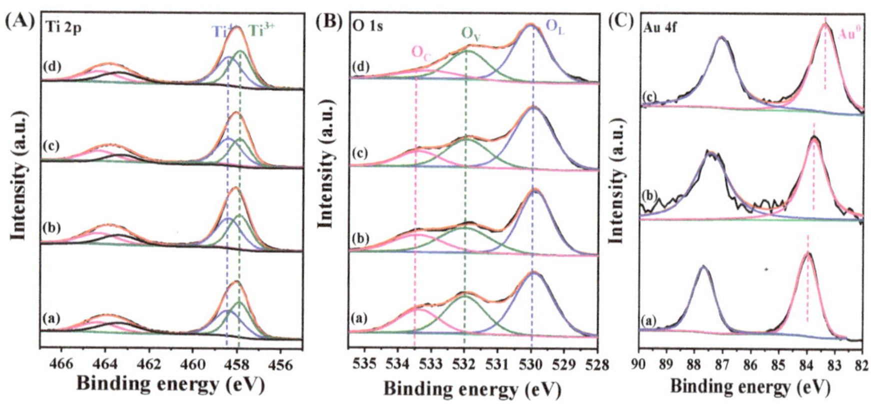

Figure 2. XPS spectra of (**A**) Ti 2p, (**B**) O 1s: (**a**) Cu_2O/TiO_2, (**b**) $Au@TiO_2/Cu_2O$, (**c**) $Au@TiO_2/Cu_2O@Au$, (**d**) $Au@Cu_2O/TiO_2$, and (**C**) Au 4f core level spectra: (**a**) $Au@TiO_2/Cu_2O$, (**b**) $Au@TiO_2/Cu_2O@Au$, (**c**) $Au@Cu_2O/TiO_2$.

Table 1. Ti 2p XPS analyses of the samples.

Sample	B.E. in $2p_{3/2}$(eV)		B.E. in $2p_{1/2}$(eV)		Ti^{3+}/Ti^{4+} Ratio [1]
	Ti^{3+}	Ti^{4+}	Ti^{3+}	Ti^{4+}	
TiO_2/Cu_2O	457.9	458.3	463.4	464.3	1.17/1
$Au@TiO_2/Cu_2O$	457.9	458.3	463.3	464.2	0.96/1
$Au@TiO_2/Cu_2O@Au$	457.9	458.3	463.4	464.3	1.03/1
$Au@Cu_2O/TiO_2$	457.8	458.2	463.4	464.2	1.11/1

[1] The value refers to the ratio of the respective integral peak area.

The O 1s spectra were deconvoluted to three peaks at 530.0 eV, 532.0 eV and 533.5 eV (Table 2), respectively, which are assigned to lattice oxygen (O_L), adsorbed oxygen adjacent to surface oxygen vacancy (O_V) and surface chemisorbed or dissociated oxygen species (O_C) (Figure 2B) [25]. The B.E. values of each oxygen species shift little among different samples. Nevertheless, the O_V/O_L ratio associated with the peak area is decreased when Au is deposited on $Au@TiO_2/Cu_2O$ and $Au@TiO_2/Cu_2O@Au$ (0.71 and 0.74) compared with that of TiO_2/Cu_2O (0.86). The O_V/O_L ratio in $Au@Cu_2O/TiO_2$ (0.81) is close to that of TiO_2/Cu_2O (0.86). This tendency is consistent with that of the Ti^{3+}/Ti^{4+} ratio. The amount of oxygen vacancy (O_V) decreases with the decreasing Ti^{3+} content because the Ti^{3+} defects are closely associated with O_V in defect-rich TiO_2 support [26].

Table 2. O 1s XPS analyses of the samples.

Sample	B.E. (eV)			O_V/O_L Ratio [1]
	O_L	O_V	O_C	
TiO_2/Cu_2O	530.0	532.0	533.5	0.86
Au@TiO_2/Cu_2O	530.0	532.0	533.5	0.71
Au@TiO_2/Cu_2O@Au	530.0	532.0	533.5	0.74
Au@Cu_2O/TiO_2	530.0	532.0	533.5	0.81

[1] The value refers to the ratio of the respective integral peak area.

Au 4f spectra of each Au-supported catalyst were shown and compared in Figure 2C. The main Au 4f peaks are located at 87.8 eV (4f $_{5/2}$) and 84.0 eV (4f $_{7/2}$). The fitted peak appears at B.E. of 84.0 eV in the Au $4f_{7/2}$ region of Au@Cu_2O/TiO_2, which is assigned to Au^0 species [27]. In contrast, when Au is deposited on both TiO_2 and Cu_2O, the B.E. shifts to a lower value of 83.7 eV, with a negative shift of 0.3 eV (Table 3). The negative shift of B.E. hints at the increased electron density of Au^0 species on Au@TiO_2/Cu_2O@Au compared to that on Au@Cu_2O/TiO_2. The B.E. continually shifts to a lower value of 83.4 eV when Au is only deposited on TiO_2. The more negative shift (−0.6 eV) indicates higher electron density of Au because of strong metal-support interaction (SMSI) between Au nanoparticles and defect-rich TiO_2 in Au@TiO_2/Cu_2O [28,29]. The negatively charged Au adjacent to Ti^{3+} sites associated with oxygen vacancies could be active sites for ethanol reforming [25]. The higher electron density of Au on Au@TiO_2/Cu_2O facilitates proton reduction to hydrogen during ethanol photoreforming.

Table 3. Au 4f XPS analyses of the samples.

Sample	B.E.(eV)	△B.E.(eV) [1]
Au@TiO2/Cu2O	83.4	−0.6
Au@TiO2/Cu2O@Au	83.7	−0.3
Au@Cu2O/TiO2	84.0	-

[1] The value refers to the shift compared to that of Au@Cu_2O/TiO_2.

XPS spectra were carried out to determine the chemical states of Cu species in the catalysts. Figure 3A shows the Cu 2p core level spectra. The binding energies at 932.1 eV and 951.9 eV, respectively, in Cu $2p_{3/2}$ and Cu $2p_{1/2}$ region were assigned to Cu^+ species [30]. The binding energies hardly shift among these catalysts. There is no shake-up satellite peak in the Cu 2p spectra. The absence of the satellite peak excludes the existence of Cu^{2+} species [31,32]. However, the Cu^+ and Cu^0 species cannot be distinguished in Cu XPS spectra because the binding energies assigned to Cu^+ and Cu^0 are very close to each other. The difference in binding energies is merely around 0.1–0.2 eV [33,34]. Consequently, Cu LMM Auger spectra were analyzed to distinguish between Cu^0 and Cu^+ species (Figure 3B). The spectra were fitted to two peaks at 917.0 eV and 919.0 eV, respectively, which is assigned to Cu^+ and Cu^0 species [35]. The Cu^+ is the dominating species on the basis of Auger spectra. The coexistence of Cu^+ and Cu^0 species in Cu_2O nanocubes or nanoparticles is also observed in the previous literatures [36]. There is no shift in the kinetic energies regardless of Au deposition on either Cu_2O or TiO_2. This is an indication that the chemical states of Cu_2O changes little in these catalysts.

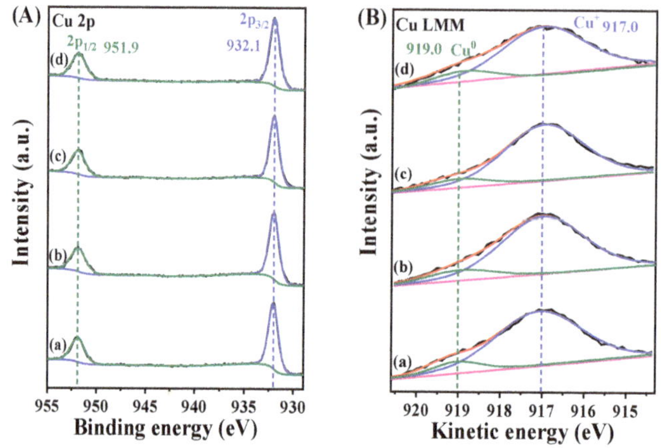

Figure 3. (**A**) Cu 2p core level spectra: (**a**) Cu_2O/TiO_2, (**b**) Au@TiO_2/Cu_2O, (**c**) Au@Cu_2O/TiO_2, (**d**) Au@TiO_2/Cu_2O@Au, and (**B**) Cu LMM Auger spectra: (**a**) Cu_2O/TiO_2, (**b**) Au@TiO_2/Cu_2O, (**c**) Au@Cu_2O/TiO_2, (**d**) Au@TiO_2/Cu_2O@Au.

The electron paramagnetic resonance (EPR) spectra are shown in Figure 4A. An intense EPR signal appears at a g value of 2.003 for Cu_2O/TiO_2. This signal is usually assigned to surface O-species, formed via the interaction of dioxygen and Ti^{3+} defects in TiO_2 [24,37–39]. The EPR signal appears at the same g value for Au@Cu_2O/TiO_2, Au@TiO_2/Cu_2O@Au and Au@TiO_2/Cu_2O. The intensity of the signal in Au-supported catalysts is lower than that in un-supported Cu_2O/TiO_2. The intensity of O-related EPR signal (g = 2.003) could be associated with the Ti^{3+}/Ti^{4+} ratio in TiO_2. The lower signal intensity in Au-supported catalysts indicates the smaller Ti^{3+}/Ti^{4+} ratio. The Au@TiO_2/Cu_2O has the smallest Ti^{3+}/Ti^{4+} ratio of 0.96 among these samples, showing the lowest signal intensity of g value. This is consistent with the results presented in XPS Ti 2p analyses (Table 1).

The digital photos of the catalysts display that all are in black color (Figure 4B). The UV-Vis absorption spectra were compared in Figure 4C. Compared to the spectrum of TiO_2, the Cu_2O/TiO_2 junction shows stronger absorption in the visible light region besides in the UV regime. The Au-supported catalysts possess more intensive absorption throughout the visible light wavelength. The absorption band (~560 nm) associated with Au surface plasmon resonance (SPR) is not clearly observed within the UV-vis absorption spectra of Au-supported catalysts because of the intense visible absorption of Cu_2O/TiO_2 itself [40]. The fluorescence emission spectra of TiO_2 and Cu_2O/TiO_2 showed a broad band peaked at 423 nm under the excitation of 350 nm (Figure 4D). The latter has lower emission intensity, indicating that the carrier recombination is inhibited due to the existence of Cu_2O/TiO_2 junction. After Au is deposited on either TiO_2 or Cu_2O, the emission intensity of supported catalysts further decreases. It indicates that forming the interface of Au and semiconducting oxides facilitates the separation of photogenerated electrons and photogenerated holes by charge transfer across the metal-oxide interface [22,41].

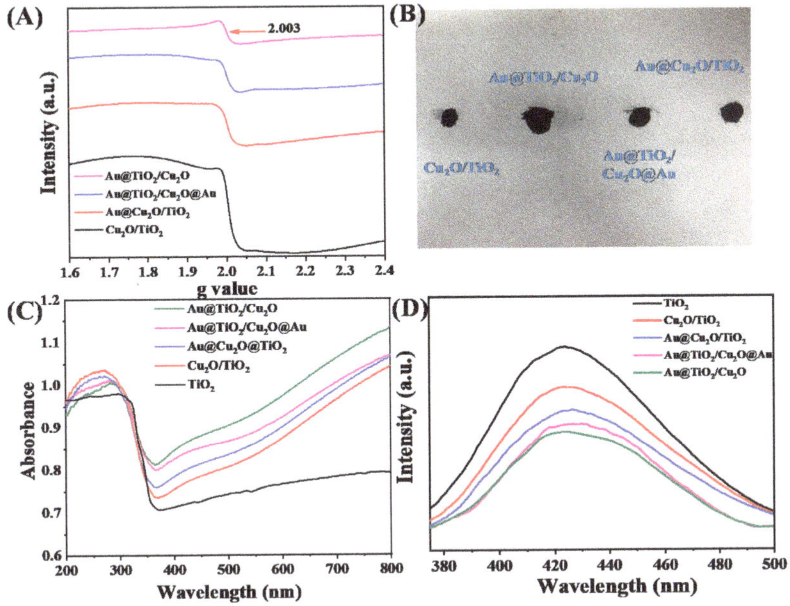

Figure 4. The EPR spectra (**A**) and digital photos (**B**) of Cu_2O/TiO_2, $Au@Cu_2O/TiO_2$, $Au@TiO_2/Cu_2O@Au$, and $Au@TiO_2/Cu_2O$; UV-Vis absorption spectra (**C**) and Fluorescence emission spectra (**D**) of TiO_2, Cu_2O/TiO_2, $Au@Cu_2O/TiO_2$, $Au@TiO_2/Cu_2O@Au$, and $Au@TiO_2/Cu_2O$. The excitation wavelength is 350 nm.

The photocurrent tests were carried out to evaluate the charge separation within the catalysts. The photocurrent–time traces of photoelectrodes were obtained in the photoelectrochemical cell at a bias of 1.23 V vs. RHE under chopped AM 1.5G light illumination (Figure 5). The Au loading on Cu_2O/TiO_2 enhances the photocurrent compared to Cu_2O/TiO_2 itself, which indicates the improved charge separation across the metal/oxide interface. The $Au@TiO_2$ electrode shows a higher photocurrent than $Au@Cu_2O/TiO_2$. The interface of Au and TiO_2 is more favorable to charge separation than that of Au and Cu_2O due to the lower conduction band potential of TiO_2 than that of Cu_2O [42]. In addition, the existence of Ti^{3+} defects associated with adjacent oxygen vacancies intensifies the interaction of Au and TiO_2 [25]. The $Au@TiO_2/Cu_2O$ electrode exhibits the highest photocurrent among all electrodes, 2.5 times higher than $Au@TiO_2$. This hints at the contribution of TiO_2/Cu_2O p-n junction combined with the interface of Au and TiO_2 to charge separation and transfer. The photocurrent of $Au@TiO_2/Cu_2O$ is also higher than $Au@TiO_2/Cu_2O/Au$. This is an indication that the site-specific deposition of Au on TiO_2 leads to charge separation more efficiently under light illumination than the random deposition of Au on both TiO_2 and Cu_2O.

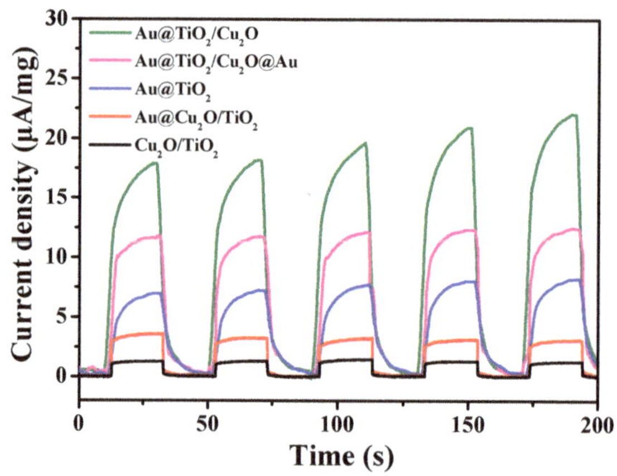

Figure 5. Photocurrent-time traces of photoelectrodes at a bias of 1.23 V vs. RHE under chopped AM 1.5G illumination (100 mW·cm^{-2}), back illumination. Solution: 0.1 M phosphate buffer (pH 7).

The photoreforming of ethanol was achieved using pure ethanol in an Ar atmosphere (1.4 bar) under simulated solar irradiation. The H_2 production rate is contrasted on each catalyst (Figure 6A). The rate of H_2 production reaches 6932 μmol g_{cat}^{-1} h^{-1} on Au@Cu_2O/TiO_2, 8.35 times higher than that on Cu_2O/TiO_2 (830 μmol g_{cat}^{-1} h^{-1}, Table 4). It is apparent that proton reduction is greatly accelerated on the Au-supported catalyst for H_2 production. Additionaly, the charge separation is improved by loading Au on the semiconductor, evidenced by photocurrent measurements (Figure 5). The Au@TiO_2 catalyst exhibits higher activity (7143 μmol g_{cat}^{-1} h^{-1}) towards H_2 production than Au@Cu_2O/TiO_2. This suggests that constructing the interface of Au and TiO_2 is more favorable than that of Au and Cu_2O, which could direct the structural design of catalysts for ethanol photoreforming. The rate of H_2 production on Au@TiO_2/Cu_2O (8548 μmol g_{cat}^{-1} h^{-1}) is apparently higher than that on either Au@TiO_2 (7143 μmol g_{cat}^{-1} h^{-1}) or Au@TiO_2/Cu_2O@Au (7348 μmol g_{cat}^{-1} h^{-1}). Table S4 compares the H_2 yield rates reported for TiO2-supported noble-metal catalysts with the result in this work. The p-n junction between TiO_2 and Cu_2O contributes the enhanced H_2 production. However, the selective deposition of Au on defect-rich TiO_2 is essential for higher activity because the Au-O_v-Ti^{3+} sites are more active in ethanol photoreforming [25]. The quantity of Au-O_v-Ti^{3+} sites is relatively less in randomly deposited Au@TiO_2/Cu_2O@Au than in site-specifically deposited Au@TiO_2/Cu_2O. As a consequence, the latter shows higher activity than the former. In addition, the lower H_2 production rate of Au@Cu_2O/TiO_2may be due to the alloying of Au-Cu, leading to a smaller number of active sites on gold according to the previous report [43].

Table 4. H_2 production via photoreforming of ethanol on the catalysts under simulated solar light.

Catalysts	Production Rate (μmol g_{cat}^{-1} h^{-1}) [1]					
	Gas Phase				Liquid Phase	
	H_2	CH_4	CO	CO_2	CH_3CHO	CH_3COOH
Au@TiO_2/Cu_2O	8548	5	-	3	8806	-
Au@Cu_2O/TiO_2	6932	12	3	9	7239	-
Au@TiO_2/Cu_2O@Au	7348	13	-	7	7561	-
Au@TiO_2	7143	7	-	7	7356	-
Cu_2O/TiO_2	830	-	-	-	1020	-
TiO_2	136	-	-	-	273	-

[1] Suspension of the photocatalyst (1 g/L) in ethanol under stirring were irradiated with simulated solar light (100 mW cm^{-2}) at Ar atmosphere (1.4 bar) at 25 °C for 6 h.

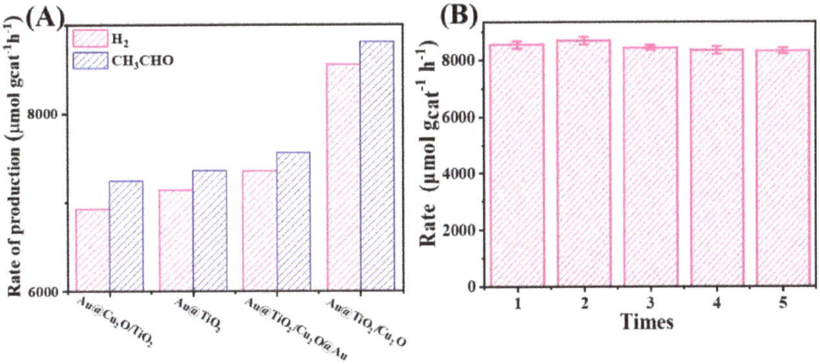

Figure 6. (**A**) H_2 and acetaldehyde production rate via the photoreforming of ethanol under simulated solar light (AM 1.5G). (**B**) The cycle tests of Au@TiO_2/Cu_2O for photocatalytic hydrogen evolution. Conditions: suspension of the photocatalyst (1 g/L) in ethanol under stirring were irradiated with simulated solar light at Ar atmosphere (1.4 bar) for 6 h.

The products except H_2 in the gas phase, e.g., CH_4, CO, CO_2, are below 0.3% on all catalysts (Table 4), which clearly indicates the efficient inhibition of C–C cleavage during the photoreforming. Also, high-purity H_2 can be produced on these catalysts. For instance, the purity of H_2 is beyond 99.9% on Au@TiO_2/Cu_2O catalyst. Acetaldehyde is the only detectable product in the liquid phase, evidenced by GC-MS analysis (Figure S2). The high selectivity of acetaldehyde is due to the absence of O_2 activation to peroxide intermediates and hydroxyl species on the surface of Au NPs under Ar ambient [43,44]. In addition, acetaldehyde is produced in almost stoichiometric yield with H_2, which is consistent with the reforming pathway of ethanol via a redox process [11]. The photo-generated holes in the valence band of semiconductor oxidize ethanol to acetaldehyde and the electrons on the active sites adjacent to Au reduce protons to H_2 during the process.

The optimal Au@TiO_2/Cu_2O catalyst was recycled five times and the H_2 yield was hardly decreased (Figure 6B). The structural stability of catalysts was further evidenced by HRTEM and Ti 2p, O 1s, Au 4f XPS and Cu LMM Auger analyses. The size of Au particles remains 2.5 nm, the same as that before the reactions (Figure S3). In addition, the Au loading after reaction is not changed by ICP measurement (0.84 wt%). The Ti^{3+}/Ti^{4+} ratio and O_V/O_L ratio in the catalyst have no change before and after reactions (Figure S4, Tables S1 and S2). The peak in the Au 4f region appears at 83.5 eV and has little shift compared to that before reactions (Figure S4, Table S3). Additionally, the peaks in Cu 2p XPS and Cu LMM Auger spectra keep unchanged after photoreforming (Figure S5). These findings verify that the catalyst is structurally stable and keeps activity after repeated use.

The in situ FTIR spectra of ethanol adsorption on the catalysts were first collected in the dark after adsorption of ethanol for 40 min, and then recorded every hour during photoreforming. The spectra of the three supported catalysts showed intensive absorption bands at 1000~1200 cm^{-1}, 1200~1500 cm^{-1}, 2700~3100 cm^{-1}, and 3600~3800 cm^{-1}, which are respectively assigned to the vibrations of ν(C–O), δ(CH_x), ν(CH_x), and ν(O–H) of ethanol (Figure 7) [45].

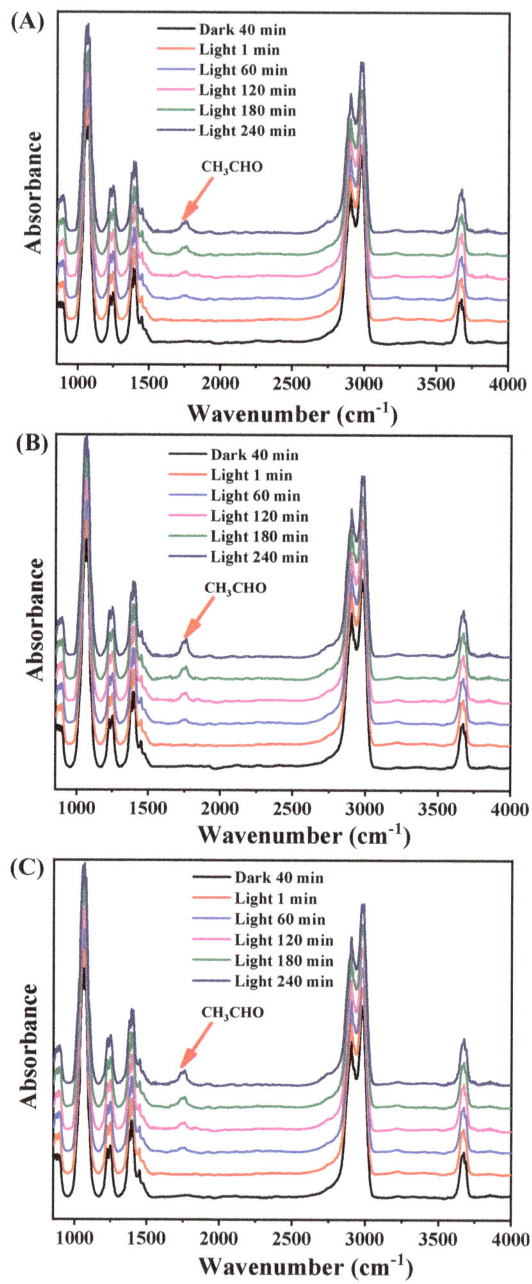

Figure 7. In situ FTIR spectra of ethanol adsorption on the catalysts during the photoreforming: (**A**) Au@Cu_2O/TiO_2, (**B**) Au@TiO_2/Cu_2O, (**C**) Au@TiO_2/Cu_2O@Au.

The band at 1755 cm^{-1} gradually increases with the extension of irradiation time, which corresponds to the vibration related with acetaldehyde molecule [46]. This is consistent with the increased acetaldehyde production in liquid phase with the reaction. The band shifts little among the three catalysts. The band at around 2350 cm^{-1}, assigned to the vibration of CO_2 [47], does not appear in the FTIR spectra. This verifies that the C–C cleavage of ethanol is greatly inhibited on the three Au-supported catalysts during the photoreforming. Consequently, the dehydrogenation of ethanol to acetaldehyde is the dominating step, accompanied with high-purity H_2 production on these catalysts.

The energy levels of semiconductors play a critical role in the activity and selectivity of ethanol photoreforming. In a catalyst with a p-n junction, the charge separation and transfer are strongly dependent on the relationships of energy levels [48]. The flat band potentials (E_{fb}) of the TiO_2 nanorods and Cu_2O nanocubes were determined by electrochemical Mott-Schottky measurements at varied frequencies and room temperature. The Mott-Schottky plots of TiO_2 exhibit positive slopes, indicative of their characteristic of n-type semiconductor (Figure 8A). The E_{fb} values are estimated by extrapolating the linear portion of the plots measured at varied frequencies to the intercept of *x*-axis. It is generally considered that the bottom of the conduction band (CB) in n-type semiconductors is approximately equal to its flat band potential. [49,50] Consequently, the conduction band of TiO_2 nanorod is −0.63 V vs. Ag/AgCl. This potential was converted to −0.02 V vs. RHE according to the following formula [51].

$$E_{RHE} = E_{Ag/AgCl} + 0.197 + 0.059pH, \tag{1}$$

The bandgap of TiO_2 nanorod was estimated to be 2.76 eV using the Kubelka–Munk equation, i.e., $F(R) = (1 - R)^2/2R$, where R is the reflectance (Figure S6). This value is smaller than that reported for bulk anatase TiO_2 of ~3.2 eV [52]. It is documented that theTi^{3+} defect-rich anatase TiO_2 exhibits narrower bandgap owing to the involvement of defect energy levels [53]. The valence band TiO_2 nanorod is calculated to be 2.74 eV from the bandgap and the conduction band level.

The Mott-Schottky plots of Cu_2O show negative slope, an indication of p-type semiconductor (Figure 8B). The flat band potential of 0.65 V vs. Ag/AgCl was converted to 1.26 V vs. RHE according to Formula (1). It is usually considered that the bottom of valence band (VB) in p-type semiconductors was 0.30 V more positive than the flat band potential [54,55]. The valence band level of Cu_2O is 1.56 V vs. RHE. Based on the bandgap of Cu_2O (2.2 eV) [56,57], the conduction band (CB) level of Cu_2O is estimated to be −0.64 V.

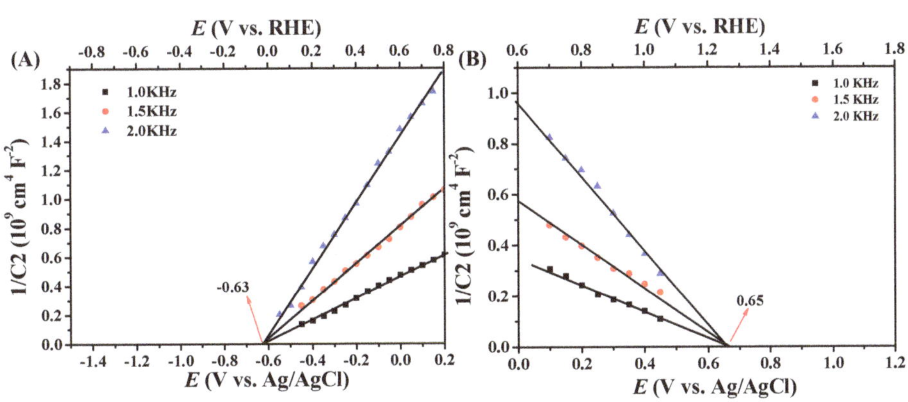

Figure 8. Mott-Schoottky plots of (**A**) TiO_2 nanorods and (**B**) Cu_2O nanocubes measured at varied frequencies and room temperature. Solution: 0.1 M phosphate buffer (pH 7).

Based on the findings above, we proposed possible charge transfer pathways on the Au@TiO_2/Cu_2O catalyst during ethanol photoreforming. The p-n junction is formed when p-type Cu_2O and n-typeTiO_2 come into contact with each other (Scheme 2). The photo-generated holes are transferred from the valence band of TiO_2 to that of Cu_2O owing to the more positive valence band level of TiO_2.

The electrons generated on Cu_2O under visible light irradiation are transferred to the conduction band of TiO_2. In addition, the hot electrons of Au under visible light irradiation are transferred to the conduction band of TiO_2 due to the SPR effect of Au. The strong interaction between Au and defect-rich TiO_2 enhances the hot electron transfer from Au to the interface of Au/TiO_2. As a consequence, the interface of Au and TiO_2 is rich in the electrons, leading to high electron density on the Au-Ov-Ti^{3+} interfacial sites. The proton reduction is promoted on the electron-rich interfacial sites, resulting in higher H_2 production rate than the Au@Cu_2O/TiO_2 catalyst without the interfacial sites. Consequently, the dual effects of Au-O_V-Ti^{3+} active sites and p-n junction endow Au@TiO_2/Cu_2O catalyst with higher activity towards ethanol photoreforming by the site-specific deposition of Au on defect-rich TiO_2 nanorods.

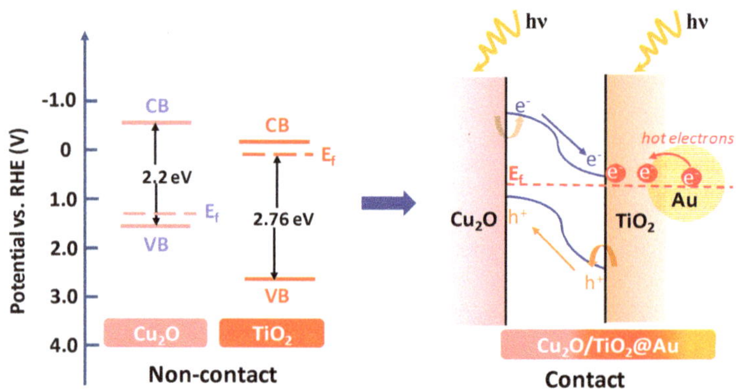

Scheme 2. Diagram of energy levels and charge transfer on Au@TiO_2/Cu_2O catalyst.

3. Materials and Methods

3.1. Materials

C_2H_5OH (anhydrous ethanol), CH_3OH (methanol), $CuSO_4$ (anhydrous copper sulfate), PVP (polyvinyl pyrrolidone), TBOT (tetrabutyltitanate), HNO_3 (nitric acid), $NaBH_4$ (sodium borohydride), AA (ascorbic acid), $HAuCl_4{\cdot}3H_2O$, EDTA (Ethylenediamine) were obtained from Shanghai Aladdin Biochemical Technology Co. Ltd. (Beijing, China). H_2 (99.9%), N_2 (99.9%) and Ar (99.9%) were supplied by Beijing Haipu Gas Co. Ltd. (Beijing, China). All reagents were of analytical grade and used as received. Deionized water was used throughout the experiments.

3.2. Synthesis of Defect-Rich TiO_2 Nanorods

The defect-rich TiO_2 nanorods were synthesized by a $NaBH_4$ reduction method according to a previous literature [22]. In a typical procedure, the mixture (TiO_2/$NaBH_4$ = 2:1 molar ratio) was continuously stirred for 30 min and treated at 350 °C for 1 h in a flow of N_2. The sample was washed with diluted HCl solution and water thoroughly. The sample was named as TiO_2.

3.3. Synthesis of Cu_2O Nanocubes

The Cu_2O nanocubes were prepared by a chemical precipitation method, modified from a literature's one [58]. In a typical run, 155 mg of $CuSO_4$ and 400 mg of PVP were dissolved in 160 mL of deionized water to obtain a light blue solution. The pH was adjusted to 10 by dropwise addition of NaOH (1 M). An amount of 176 mg of ascorbic acid (AA) was dispersed in the solution to reduce Cu^{2+} to Cu_2O. The suspension was stirred for 20 min. Subsequently, the suspension was centrifuged and washed thoroughly with deionized water and ethanol to remove PVP. The product was collected after the color of solution changed from light blue to brownish yellow. The solid product was finally dried under vacuum at 60 °C.

3.4. Synthesis of Cu_2O/TiO_2

2 g of TiO_2 nanorod sample was added into the $CuSO_4$ and PVP solution, and the pH of the solution was adjusted to 10 with an aqueous solution of NaOH (1 M). Subsequently, AA was added to the solution and stirred for 20 min. The solid product was collected by centrifugation and washed with ethanol and water. The product was finally dried under vacuum at 60 °C.

3.5. Synthesis of Au@TiO_2 and Au@TiO_2/Cu_2O

2 g of TiO_2 nanorod sample was dispersed in 200 mL of H_2O containing 2 mL of $HAuCl_4$ solution (0.051 M). The pH of suspension was adjusted to 10 with an aqueous solution of NaOH (1 M). 10 mL of methanol was added and stirred for 3 h. The Au@TiO_2 sample was obtained by a photo-reduction method under UV irradiation. After the photo-reduction, the precipitation was centrifuged, washed with water thoroughly to neutral and dried in vacuum at 60 °C. The Au@TiO_2 product was suspended in 160 mL of aqueous solution containing 155 mg of $CuSO_4$ and 400 mg of PVP. The pH was adjusted to 10 with an aqueous solution of NaOH (1 M). 176 mg of AA was added and stirred for 20 min. The solid product was collected by centrifugation and washed with by ethanol and water and finally dried under vacuum at 60 °C. The product was denoted as Au@TiO_2/Cu_2O.

3.6. Synthesis of Au@Cu_2O/TiO_2

The as-synthesized Cu_2O and PVP was suspended in 100 mL of H_2O. 2 mL of $HAuCl_4$ solution (0.051 M) was added and the pH of suspension was adjusted to 10 with an aqueous solution of NaOH (1 M). 10 mL of methanol was added and stirred for 3 h. The Au@Cu_2O sample was obtained by a photo-reduction method under visible light irradiation. The as-synthesized TiO_2 nanorod was suspended in the solution and stirred for 1 h. The solid product was collected by centrifugation and washed with by ethanol and water and finally dried under vacuum at 60 °C. The product was denoted as Au@Cu_2O/TiO_2.

3.7. Synthesis of Au@TiO_2/Cu_2O@Au

2 g Cu_2O/TiO_2 sample was suspended in 200 mL of H_2O containing 2 mL of $HAuCl_4$ solution (0.051 M). The pH of suspension was adjusted to 10 with an aqueous solution of NaOH (1 M). 10 mL of methanol was added and stirred for 3 h. The Au@TiO_2/Cu_2O@Au sample was obtained by a photo-reduction method under UV-visible light irradiation. The solid product was collected by centrifugation and washed with ethanol and water and finally dried under vacuum at 60 °C.

3.8. Characterization

Powder X-ray diffraction (XRD) patterns of the samples were obtained by a Shimadzu XRD-6000 diffractormeter using graphite-filtered CuKα radiation (40 kV, 30 mA, λ = 0.15418 nm) in a 2theta range of 10–70°.

Transmission electron microscopy (TEM) measurements were carried out on a JEOL JEM-3010 high-resolution transmission electron microscope.

Elemental analysis was performed on a Shimadzu ICPS-7500 inductively coupled plasma atomic emission spectrometer (ICP-AES).

The XPS spectra were recorded on a Thermo VG ESCALAB MK II X-ray photoelectron spectrometer at a pressure of 2×10^{-9} Pa using Al Kα X-ray as the excitation source (1486.6 eV). The positions of all binding energies were calibrated using theC1s line at 284.6 eV.

UV-visible diffuse reflectance spectra were performed on a Shimadzu UV-3000 spectrometer equipped with an integrating sphere attachment with $BaSO_4$ (10 mg) as reference.

Photoluminescence (PL) emission spectra were recorded by a Hitachi F-7000 spectrofluorometer using laser excitation at a wavelength of 350 nm.

Electron paramagnetic resonance (EPR) measurements were performed on a Bruker E500 spectrometer with a 9.53 GHz X-band. The sample mass was 50 mg. The spectra were recorded in the magnet field range of 318–328 mT.

The liquid phase products were determined by GC-MS measurements using a Thermo Fisher ISQ Trace1300 and the sample volume was 1 μL.

Electrochemical impedance spectroscopy (EIS) was performed on an electrochemical workstation at an open-circuit voltage of 0.3 V vs. RHE under illumination with 10 mV amplitude of perturbation and a frequency between 1.0 kHz, 1.5 kHz and 2.0 kHz. Mott-Schottky plots were measured at room temperature in the dark.

In situ Fourier Transform Infrared Spectroscopy (FTIR) was conducted in an in situ reaction cell on a Bruker Tensor II spectrometer installed with MCT narrow-band detector. The sample was pretreated in a flow of high-purity N_2 at 100 °C for 1 h. After an initial scan as the background spectrum, ethanol was induced into the cell through a flow of N_2 for 30 min. After flowing N_2 to remove the residual ethanol vapor, the FTIR spectra were collected in the range of 4000~950 cm^{-1} at room temperature.

3.9. Photoreforming of Ethanol

In a typical run, 25 mg of catalyst was added to 25 mL of ethanol and ultrasonically dispersed for 15 min. The suspension was transferred into a high-pressure stainless steel reactor (volume: 50 mL) equipped with a sapphire crystal window. A flow of Ar was purged into the reactor for 30 min and the reactor was evacuated. The Ar was re-purged and the pressure in the reactor was maintained at 1.4 bar. The suspension was irradiated by a 300 W Xenon lamp equipped with an AM 1.5G filter (100 $mW{\cdot}cm^{-2}$) under magnetic stirring for 6 h. After the reaction, the catalyst was removed from the solution by filtration. The gaseous products were detected by an online gas chromatograph (GC-2014C, Shimadzu, Japan) equipped with a high-sensitivity thermal conductivity detector (TCD) and Ar was used as the carrier gas. The liquid products were analyzed by a gas chromatograph (GC-2014C, Shimadzu) equipped with a flame ionization detector (FID). The reaction rate of H_2 followed the formula below:

$$\text{Reaction rate of production} = \frac{n_{\text{production}}(\mu\text{mol})}{m_{\text{cat.}}(\text{g}) \times \text{time (h)}} \tag{2}$$

3.10. Recycled Use

The catalyst was separated from the solution after the reaction, washed with deionized water, and finally dried under vacuum at 60 °C for 6 h. The dried catalyst was reused in a next catalysis run under the same reaction conditions.

4. Conclusions

The TiO_2/Cu_2O-supported Au catalysts were delicately constructed by the site-specific deposition of Au on either defect-rich TiO_2 nanorods or Cu_2O nanocubes. The selective anchoring of Au nanoparticles on TiO_2 nanorods combined with the p-n junction of TiO_2/Cu_2O leads to the highest activity towards ethanol photoreforming. The H_2 production rate reaches a record level of 8548 μmol g_{cat}^{-1} h^{-1} under simulated solar light. The acetaldehyde in liquid phase was generated at an almost stoichiometric rate, which indicates the effective inhibition of C−C cleavage of ethanol to CH_4 or CO_2. Extensive spectroscopic studies verified that Au species adjacent to Ti^{3+} defects and the associated oxygen vacancies on TiO_2 nanorods activate the proton reduction to H_2. The p-n junction between TiO_2 and Cu_2O facilitates charge separation and transfer owing to the matching of energy levels, which accelerates photo-generated hole transfer and the dehydrogenation of ethanol to acetaldehyde on the junction. This site-specific deposition strategy could be applicable in the enhancement in other biomass hydrogen production by rationally designing the delicate structure of catalysts and maximizing the catalytic capability.

Supplementary Materials: The following are available online at http://www.mdpi.com/2073-4344/10/5/539/s1, Figure S1: XRD patterns; Figure S2: GC-MS spectra; Figure S3: HRTEM photographs; Figure S4: Ti 2p, O 1s and Au 4f XPS spectra; Figure S5: Cu 2p XPS and Cu LMM Auger spectra; Figure S6: Bandgap estimation; Table S1: Ti 2p XPS analyses; Table S2: O 1s XPS analyses; Table S3: Au 4f XPS analyses; Table S4: Comparisons of photocatalysts for ethanol photoreforming.

Author Contributions: Conceptualization, X.X.; methodology, T.Z. and X.Z.; formal analysis, L.L., T.Z., X.X.; investigation, X.Z., L.L.; data curation, L.L., T.Z., R.Y.; writing—original draft preparation, L.L., T.Z., X.X.; writing—review and editing, T.Z., R.Y., Y.L., B.Z., X.X.; supervision, X.X. All authors have read and agreed to the published version of the manuscript.

Funding: This research was funded by the National Natural Science Foundation of China (grant 21978021, 21521005), the National Key R&D Program of China (grant 2017YFA0206804), and the Fundamental Research Funds for the Central Universities.

Conflicts of Interest: The authors declare no competing financial interest.

References

1. Lu, X.H.; Xie, S.L.; Yang, H.; Tong, Y.X.; Ji, H.B. Photoelectrochemical hydrogen production from biomass derivatives and water. *Chem. Soc. Rev.* **2014**, *43*, 7581–7593. [CrossRef] [PubMed]
2. Zhao, X.X.; Feng, J.R.; Liu, J.W.; Lu, J.; Shi, W.; Yang, G.M.; Wang, G.C.; Feng, P.Y.; Cheng, P. Metal-Organic Framework-Derived ZnO/ZnS Heteronano structures for Efficient Visible Light Driven Photocatalytic Hydrogen Production. *Adv. Sci.* **2018**, *5*, 1700590. [CrossRef] [PubMed]
3. Zhao, X.X.; Feng, J.R.; Liu, J.; Shi, W.; Yang, G.M.; Wang, G.C.; Cheng, P. An Efficient, Visible Light Driven, Hydrogen Evolution Catalyst NiS/$Zn_xCd_{1-x}S$ Nanocrystal Derived from a Metal–Organic Framework. *Angew. Chem. Int. Ed.* **2018**, *57*, 9790. [CrossRef] [PubMed]
4. Shimura, K.; Yoshida, H. Heterogeneous photocatalytic hydrogen production from water and biomass derivatives. *Energy Environ. Sci.* **2011**, *4*, 2467–2484. [CrossRef]
5. Hou, T.F.; Zhang, S.Y.; Chen, Y.D.; Wang, D.Z.; Cai, W.J. Hydrogen production from ethanol reforming: Catalysts and reaction mechanism. *Renew. Sustain. Energy Rev.* **2015**, *44*, 132–148. [CrossRef]
6. Yamazaki, T.; Kikuchi, N.; Katohb, M.; Hirosea, T.; Saitoc, H.; Yoshikawa, T.; Wada, M. Behavior of steam reforming reaction for bio-ethanol over Pt/ZrO_2 catalysts. *Appl. Catal. B Environ.* **2010**, *99*, 81–88. [CrossRef]
7. Gasparotto, A.; Barreca, D.; Bekermann, D.; Devi, A.; Fischer, R.A.; Fornasiero, P.; Gombac, V.; Lebedev, O.I.; Maccato, C.; Montini, T.; et al. F-Doped Co_3O_4 Photocatalysts for Sustainable H_2 Generation from Water/Ethanol. *J. Am. Chem. Soc.* **2011**, *133*, 19362–19365. [CrossRef]
8. Bac, S.; Keskin, S.; Avci, A.K. Recent advances in materials for high purity H_2 production by ethanol and glycerol steam reforming. *Int. J. Hydrogen Energy* **2019**. [CrossRef]
9. Llorca, J.; Homs, N.; Sales, J.; Fierro, J.L.G.; Piscina, P.R.D.L. Effect of sodium addition on the performance of Co-ZnO-based catalysts for hydrogen production from bioethanol. *J. Catal.* **2004**, *222*, 470–480. [CrossRef]
10. Ciambelli, P.; Palma, V.; Ruggiero, A. Low temperature catalytic steam reforming of ethanol. The effect of the support on the activity and stability of Pt catalysts. *Appl. Catal. B Environ.* **2010**, *96*, 18–27. [CrossRef]
11. Puga, A.V.; Forneli, A.; García, H.; Corma, A. Production of H_2 by Ethanol Photoreforming on Au/TiO_2. *Adv. Funct. Mater.* **2014**, *24*, 241–248. [CrossRef]
12. Mizukoshi, Y.; Sato, K.; Konno, T.J.; Masahashi, N. Dependence of photocatalytic activities upon the structures of Au/Pd bimetallic nanoparticles immobilized on TiO_2 surface. *Appl. Catal. B Environ.* **2010**, *94*, 248–253. [CrossRef]
13. Strataki, N.; Bekiari, V.; Kondarides, D.I.; Lianos, P. Hydrogen production by photocatalytic alcohol reforming employing highly efficient nanocrystalline titania films. *Appl. Catal. B Environ.* **2007**, *77*, 184–189. [CrossRef]
14. Lou, Z.; Fujitsuka, M.; Majima, T. Pt–Au triangular nanoprisms with strong dipole plasmon resonance for hydrogen generation studied by single-particle spectroscopy. *ACS Nano* **2016**, *10*, 6299–6305. [CrossRef] [PubMed]

15. Wu, B.H.; Liu, D.Y.; Mubeen, S.; Chuong, T.T.; Moskovits, M.; Stucky, G.D. Anisotropic growth of TiO_2 onto gold nanorods for plasmon-enhanced hydrogen productionfrom water reduction. *J. Am. Chem. Soc.* **2016**, *138*, 1114–1117. [CrossRef]
16. Chang, A.; Peng, W.S.; Tsaia, I.-T.; Chiang, L.F.; Yang, C.M. Efficient hydrogen production by selective alcohol photoreforming on plasmonic photocatalyst comprising sandwiched Au nanodisks and TiO_2. *Appl. Catal. B Environ.* **2019**, *255*, 117773. [CrossRef]
17. Nadeem, M.A.; Murdoch, M.; Waterhouse, G.I.N.; Metson, J.B.; Keane, M.A.; Llorca, J.; Idriss, H. Photoreaction of ethanol on Au/TiO_2 anatase: Comparing the micro to nanoparticle size activities of the support for hydrogen production. *J. Photochem. Photobiol. A Chem.* **2010**, *216*, 250–255. [CrossRef]
18. Shiraishi, Y.; Imai, J.; Yasumoto, N.; Sakamoto, H.; Tanaka, S.; Ichikawa, S.; Hirai, T. Doping of Nb^{5+} Species at the Au-TiO_2 Interface for Plasmonic Photocatalysis Enhancement. *Langmuir* **2019**, *35*, 5455–5462. [CrossRef]
19. Majhi, S.M.; Rai, P.; Raj, S.; Chon, B.S.; Park, K.K.; Yu, Y.T. Effect of Au Nanorods on Potential Barrier Modulation in Morphologically Controlled Au@Cu_2O Core−Shell Nanoreactors for Gas Sensor Applications. *ACS Appl. Mater. Interfaces* **2014**, *6*, 7491–7497. [CrossRef]
20. Li, G.J.; Huang, J.Q.; Chen, J.; Deng, Z.J.; Huang, Q.F.; Liu, Z.Q.; Guo, W.; Cao, R. Highly Active Photocatalyst of Cu_2O/TiO_2 Octahedron for Hydrogen Generation. *ACS Omega* **2019**, *4*, 3392–3397. [CrossRef]
21. Lu, C.L.; Prasad, K.S.; Wu, H.L.; Ho, J.A.; Huang, M.H. Au Nanocube-Directed Fabrication of Au-Pd Core-Shell Nanocrystals with Tetrahexahedral, Concave Octahedral, and Octahedral Structures and Their Electrocatalytic Activity. *J. Am. Chem. Soc.* **2010**, *132*, 14546–14553. [CrossRef] [PubMed]
22. Zhang, Y.; Xing, Z.; Liu, X.F.; Li, Z.Z.; Wu, X.Y.; Jiang, J.J.; Li, M.; Zhu, Q.; Zhou, W. Ti^{3+} Self-Doped Blue TiO_2(B) Single Crystalline Nanorods for Efficient Solar-Driven Photocatalytic Performance. *ACS Appl. Mater. Interfaces* **2016**, *8*, 26851–26859. [CrossRef] [PubMed]
23. Li, G.; Li, J.; Li, G.; Jiang, G. N and Ti^{3+} co-doped 3D anatase TiO_2 superstructures composed of ultrathin nanosheets with enhanced visible light photocatalytic activity. *J. Mater. Chem. A* **2015**, *3*, 22073–22080. [CrossRef]
24. Wu, S.M.; Liu, X.L.; Lian, X.L.; Tian, G.; Janiak, C.; Zhang, Y.X.; Lu, Y.; Yu, H.Z.; Hu, J.; Wei, H.; et al. Homojunction of Oxygen and Titanium Vacancies and its Interfacial n−p Effect. *Adv. Mater.* **2018**, *30*, 1802173–1802180. [CrossRef] [PubMed]
25. Zhang, X.; Luo, L.; Yun, R.P.; Pu, M.; Zhang, B.; Xiang, X. Increasing the Activity and Selectivity of TiO_2 Supported Au Catalysts for Renewable Hydrogen Generation from Ethanol Photoreforming by Engineering Ti^{3+} Defects. *ACS Sustain. Chem. Eng.* **2019**, *7*, 13856–13864. [CrossRef]
26. Wan, J.W.; Chen, W.X.; Jia, C.Y.; Zheng, L.R.; Dong, J.C.; Zheng, X.S.; Wang, Y.; Yan, W.S.; Chen, C.; Peng, Q.; et al. Defect Effects on TiO_2 Nanosheets: Stabilizing Single Atomic Site Au and Promoting Catalytic Properties. *Adv. Mater.* **2018**, *30*, 1705369–1705377. [CrossRef]
27. Wang, Y.; Widmann, D.; Behm, R.J. Influence of TiO_2 Bulk Defects on CO Adsorption and CO Oxidation on Au/TiO_2: Electronic Metal-Support Interactions (EMSIs) in Supported Au Catalysts. *ACS Catal.* **2017**, *7*, 2339–2345. [CrossRef]
28. Liu, N.; Xu, M.; Yang, Y.; Zhang, S.; Zhang, J.; Wang, W.; Zheng, L.; Hong, S.; Wei, M. $Au^{\delta-}$–Ov–Ti^{3+} Interfacial Site: Catalytic Active Center towards Low-Temperature Water Gas Shift Reaction. *ACS Catal.* **2019**, *9*, 2707–2717. [CrossRef]
29. Li, H.X.; Bian, Z.F.; Zhu, J.; Huo, Y.N.; Huo, H.; Lu, Y.F. Mesoporous Au/TiO_2 Nanocomposites with Enhanced Photocatalytic Activity. *J. Am. Chem. Soc.* **2007**, *129*, 4538–4539. [CrossRef]
30. Zhu, S.Y.; Liang, S.Y.; Tong, Y.C.; An, X.H.; Long, J.L.; Fu, X.Z.; Wang, X.X. Photocatalytic reduction of CO_2 with H_2O to CH_4 on Cu(I) supported TiO_2 nanosheets with defective {001} facets. *Phys. Chem. Chem. Phys.* **2015**, *17*, 9761–9770. [CrossRef]
31. Babu, S.G.; Vinoth, R.; Kumar, D.P.; Shankar, M.V.; Chou, H.L.; Vinodgopal, K.; Neppolian, B. Influence of electron storing, transferring and shuttling assets of reduced graphene oxide at the interfacial copper doped TiO_2 p-n heterojunction for increased hydrogen production. *Nanoscale* **2015**, *7*, 7849–7857. [CrossRef] [PubMed]
32. Gopiraman, M.; Babu, S.G.; Khatri, Z.; Kai, W.; Kim, Y.A.; Endo, M.; Karvembu, R.; Kim, I.S. An efficient, reusable copperoxide/carbon-nanotube catalyst for N-arylation of imidazole. *Carbon* **2013**, *62*, 135–148. [CrossRef]

33. Movahed, S.K.; Dabiri, M.; Bazgir, A. A one-step method for preparation of Cu@Cu_2O nanoparticles on reduced graphene oxide and their catalytic activities in N-arylation of N-heterocycles. *Appl. Catal. A* **2014**, *481*, 79–88. [CrossRef]
34. Balasubramanian, P.; Balamurugan, T.S.T.; Chen, S.M.; Chen, T.W.; Sathesh, T. Rational design of Cu@Cu_2O nanospheres anchored B, N co-doped mesoporous carbon: A sustainable electrocatalyst to assay eminent neurotransmitters Acetylcholine and Dopamine. *ACS Sustain. Chem. Eng.* **2019**, *7*, 5669–5680. [CrossRef]
35. Mcshane, C.M.; Choi, K.S. Junction studies on electrochemically fabricated p–n Cu_2O homojunction solar cells for efficiency enhancement. *Phys. Chem. Chem. Phys.* **2012**, *14*, 6112–6118. [CrossRef]
36. Xi, Z.H.; Li, C.J.; Zhang, L.; Xing, M.Y.; Zhang, J.L. Synergistic effect of Cu_2O/TiO_2 heterostructure nanoparticle and its high H_2 evolution activity. *Int. J. Hydrogen Energy* **2014**, *39*, 6345–6353. [CrossRef]
37. Yu, M.; Kim, B.; Kim, Y.K. Highly Enhanced Photoactivity of Anatase TiO_2 Nanocrystals by Controlled Hydrogenation-Induced Surface Defects. *ACS Catal.* **2013**, *3*, 2479–2486. [CrossRef]
38. Wang, W.; Lu, C.H.; Ni, Y.R.; Su, M.X.; Xu, Z.Z. A new sight on hydrogenation of F and N-F doped {001} facets dominated anatase TiO_2 for efficient visible light photocatalyst. *Appl. Catal. B Environ.* **2012**, *127*, 28–35. [CrossRef]
39. Zuo, F.; Wang, L.; Wu, T.; Zhang, Z.; Borchardt, D.; Feng, P. Self-Doped Ti^{3+} Enhanced Photocatalyst for Hydrogen Productionunder Visible Light. *J. Am. Chem. Soc.* **2010**, *132*, 11856–11857. [CrossRef]
40. Chen, J.J.; Wu, J.C.S.; Wu, P.C.; Tsai, D.P. Plasmonic Photocatalyst for H_2 Evolution in Photocatalytic Water Splitting. *J. Phys. Chem. C* **2011**, *115*, 210–216. [CrossRef]
41. Qian, K.; Sweeny, B.C.; Johnston-Peck, A.C.; Niu, W.X.; Graham, J.O.; DuChene, J.S.; Qiu, J.J.; Wang, Y.C.; Engelhard, M.H.; Su, D.; et al. Surface Plasmon-Driven Water Reduction: Gold Nanoparticle Size Matters. *J. Am. Chem. Soc.* **2014**, *136*, 9842–9845. [CrossRef] [PubMed]
42. Zhu, Y.C.; Liu, Y.L.; Xu, Y.T.; Ruan, Y.F.; Fan, G.C.; Zhao, W.W.; Xu, J.J.; Chen, H.Y. Three-Dimensional TiO_2@Cu_2O@Nickel Foam Electrode: Design, Characterization, and Validation of O_2-Independent Photocathodic Enzymatic Bioanalysis. *ACS Appl. Mater. Interfaces* **2019**, *11*, 25702–25707. [CrossRef] [PubMed]
43. Hua, W.J.; Li, D.P.; Yang, Y.; Li, T.; Chen, H.P.; Liu, P. Copper ferrite supported gold nanoparticles as efficient and recyclable catalyst for liquid-phase ethanol oxidation. *J. Catal.* **2018**, *357*, 108–117. [CrossRef]
44. Wolkenstein, T. Theory of the photoadsorption effect in semiconductors. *Prog. Surf. Sci.* **1975**, *6*, 213–240. [CrossRef]
45. Gharahshiran, V.S.; Yousefpour, M. Synthesis and characterization of Zr-promoted Ni-Co bimetallic catalyst supported OMC and investigation of its catalytic performance in steam reforming of ethanol. *Int. J. Hydrogen Energy* **2018**, *43*, 7020–7037. [CrossRef]
46. Mironova, Y.; Lytkina, A.A.; Ermilova, M.M.; Efimov, M.N.; Zemtsov, L.M.; Orekhova, N.V.; Karpacheva, G.P.; Bondarenko, G.N.; Muraviev, D.N.; Yaroslavtsev, A.B. Ethanol and methanol steam reforming on transition metal catalysts supported on detonation synthesis nanodiamonds for hydrogen production. *Int. J. Hydrogen Energy* **2015**, *40*, 3557–3565. [CrossRef]
47. Mulewa, W.; Tahir, M.; Amin, N.A.S. MMT-supported Ni/TiO_2 nanocomposite for low temperature ethanol steam reforming toward hydrogen production. *Chem. Eng. J.* **2017**, *326*, 956–969. [CrossRef]
48. Liu, H.; Zhu, X.D.; Han, R.; Dai, Y.X.; Sun, Y.L.; Lin, Y.N.; Gao, D.D.; Wang, X.Y.; Lu, C.N. Study on the internal electric field in the Cu_2O/g-C_3N_4 p-n heterojunction structure for enhancing visible light photocatalytic activity. *New J. Chem.* **2020**. [CrossRef]
49. Zhang, Z.; Long, J.; Yang, L.; Chen, W.; Dai, W.; Fu, X.; Wang, X. Organic semiconductor for artificial photosynthesis: Water splitting into hydrogen by a bioinspired $C_3N_3S_3$ polymer under visible light irradiation. *Chem. Sci.* **2011**, *2*, 1826–1830. [CrossRef]
50. Maeda, K.; Sekizawa, K.; Ishitani, O. A polymeric-semiconductor–metal-complex hybrid photocatalyst for visible-light CO_2 reduction. *Chem. Commun.* **2013**, *49*, 10127–10129. [CrossRef]
51. He, W.H.; Wang, R.R.; Zhang, L.; Zhu, J.; Xiang, X.; Li, F. Enhanced photoelectrochemical water oxidation on a $BiVO_4$ photoanode modified with multifunctional layered double hydroxide nanowalls. *J. Mater. Chem. A* **2015**, *3*, 17977–17983. [CrossRef]
52. Ramírez-Ortega, A.M.; Melendez, P.; Acevedo-Pena, I.; Gonzalez, R. Arroyo, Semiconducting properties of ZnO/TiO_2 composites by electrochemical measurements and their relationship with photocatalytic activity. *Electrochim. Acta* **2014**, *140*, 541–549. [CrossRef]

53. Hu, Y.; Zhou, W.; Zhang, K.F.; Zhang, X.C.; Wang, L.; Jiang, B.J.; Tian, G.H.; Zhao, D.Y.; Fu, H.G. Facile strategy for controllable synthesis of stable mesoporous black TiO_2 hollow spheres with efficient solar-driven photocatalytic hydrogen evolution. *J. Mater. Chem. A* **2016**, *4*, 7495–7502. [CrossRef]
54. Matsumoto, Y. Energy Positions of Oxide Semiconductors and Photocatalysis with Iron Complex Oxides. *J. Solid State Chem.* **1996**, *126*, 227–234. [CrossRef]
55. Matsumoto, Y.; Sugiyama, K.; Sato, E. Improvement of $CaFe_2O_4$ Photocathode by doping with Na and Mg. *J. Solid State Chem.* **1988**, *74*, 117–125. [CrossRef]
56. Zhang, J.; Liu, J.; Peng, Q.; Wang, X.; Li, Y. Nearly Monodisperse Cu_2O and CuO Nanospheres: Preparation and Applications for Sensitive Gas Sensors. *Chem. Mater.* **2006**, *18*, 867–871. [CrossRef]
57. Musa, A.O.; Akomolafe, T.; Carter, M.J. Production of cuprous oxide, a solar cell material, by thermal oxidation and a study of its physical and electrical properties. *Sol. Energy Mater. Sol. C* **1998**, *51*, 305–316. [CrossRef]
58. Zhang, D.-F.; Zhang, H.; Guo, L.; Zheng, K.; Han, X.-D.; Zhang, Z. Delicate control of crystallographic facet-oriented Cu_2O nanocrystals and the correlated adsorption ability. *J. Mater. Chem.* **2009**, *19*, 5220–5225. [CrossRef]

catalysts

MDPI

Article

A 3D Hierarchical Pancake-Like Porous Carbon Nitride for Highly Enhanced Visible-Light Photocatalytic H_2 Evolution

Xiaobin Qiu, Lingfang Qiu *, Mengfan Ma, Yingying Hou and Shuwang Duo *

Jiangxi Key Laboratory of urfacSe Engineering, Jiangxi Science and Technology Normal University, Nanchang 330013, China; xbluyin@126.com (X.Q.); MF1783975@126.com (M.M.); hyy18170401838@163.com (Y.H.)

* Correspondence: qlf1108@163.com (L.Q.); swduo@imr.ac.cn (S.D.)

Received: 26 November 2019; Accepted: 2 January 2020; Published: 4 January 2020

Abstract: Polymeric carbon nitride is a fascinating visible-light-response metal-free semiconductor photocatalyst in recent decades. Nevertheless, the photocatalytic H_2 efficiency is unsatisfactory due to the insufficient visible-light harvesting capacity and low quantum yields caused by the bulky structure seriously limited its applications. To overcome these defects, in this research, a 3D hierarchical pancake-like porous carbon nitride (PPCN) was successfully fabricated by a facile bottom-up method. The as-prepared photocatalyst exhibit enlarged surface area, enriched reactive sites, improved charge carrier transformation and separation efficiency, and expanded bandgap with a more negative conduction band towardan enhanced reduction ability. All these features synergistically enhanced the photocatalytic H_2 evolution efficiency of 3% Pt@PPCN (430 μmol g^{-1} h^{-1}) under the visible light illumination ($\lambda \geq 420$ nm), which was nine-fold higher than that of 3% Pt@bulk C_3N_4 (BCN) (45 μmol g^{-1} h^{-1}). The improved structure and enhanced photoelectric properties were systematically investigated by different characterization techniques. This research may provide an insightful synthesis strategy for polymeric carbon nitride with excellent light-harvesting capacity and enhanced separation of charges toward remarkable photocatalytic H_2 for water splitting.

Keywords: pancake-like porous carbon nitride; bottom-up method; water splitting; visible light photocatalyst

1. Introduction

With fossil fuel reserves dwindling every day, energy shortage and environmental pollution issues have become increasingly prominent. Developing clean and renewable energy sources is urgent to meet the sustained growth in energy demand [1–6]. Since the discovery of the Honda–Fujishima effect in 1972 [7], photocatalytic hydrogen evolution reaction (HER) through water splitting has gained considerable attention due to its enormous potential to convert renewable solar energy to clean sustainable energy [8–12]. Researchers have devoted a significant amount of effort to investigate more effective photocatalytic systems for hydrogen evolution via photocatalytic water splitting [13–16]. Among multitudinous proposed semiconductor photocatalysts, polymeric carbon nitride (CN) has been considered promising for visible-light driven hydrogen evolution due to its tunable bandgap, low cost, earth-abundant nature, nontoxicity, facile preparation, being environmentally benign, and its remarkable physical and chemical stability [17–22]. Unfortunately, the photocatalytic performance of bulky CN was severely limited, resulting from the small specific surface area, large recombination rates of photo-excited electron-hole pairs, weak light-harvesting capacity, and inadequate catalytic active sites [23–28]. To address these shortcomings, many strategies have been developed, such as heteroatom doping [29,30], electronic structure modulation [31,32], nano-structuring [33,34], dye sensitization [35–37], and heterojunction construction [38–41]. Among them, nano-structuring is considered one of the most efficient and accessible strategies, as the enlarged specific surface area simultaneously abundant and quality active sites are not only favorable to the transfer of photo-excited charge carriers but also shortened the diffusion length for both ions and electrons [42,43]. On the other hand, the nano-structuring designed materials commonly exhibit the quantum confinement effect [24] with an enlarged electronic bandgap would be favorable to an enhanced redox ability toward photocatalytic reaction, promoting the quality and quantity of photo-induced charge carriers to some extent.

The nanostructural CN can be achieved by many methods. template-assistance strategies, top-down methods, and hydrothermal approaches have been regarded as representative and impelling strategies to fabricate multidimensional CN [44,45]. Nevertheless, these methods have many unsatisfactory deficiencies, such as a shortage of template species, large time consumption, and being environmentally unfriendly, resulting from the introduction of hazardous reagents, challenging the removal of the byproducts [46,47]. In contrast, the highly efficient photocatalysis materials benefited from the improved structure and accelerated the separation efficiency of photo-induced charge carriers, which is relatively convenient to obtain by a facile bottom-up method [48,49].

Porous carbon nitride as a photocatalyst was introduced by Kang et al. Nevertheless, the material showed undesirable photocatalytic performance with a poor quantum efficiency of ~0.5%. Rahman et al. improved it to some extent, which reached a benchmark quantum efficiency of ~6% [50,51]. The synthesized photocatalyst exhibited similar structure compared to that synthesized by Rahmanetal. While the PPCN presented a higher quantum efficiency of ~19%.In this work, we report a low-cost and straightforward bottom-up method for the production of hierarchical pancake-like porous carbon nitride, the typical synthesis procedure is presented in Scheme 1. Typically, the cyanuric acid–melamine complex, which is suitable for controlling molecular self-assembly [52], was firstly synthesized. Then, ethanol and glycerol intercalated into the framework of the precursor. Next, a thermal-induced exfoliation and polycondensation process was performed to obtain the resultant products. The as-synthesized PPCN exhibited unique hierarchical pancake-like porous morphology structure and remarkable photoelectric properties. As expected, by virtue of an enlarged specific surface area up to 79.663 m^2g^{-1}, enhancing the light-harvesting capacity, and enriching the reactive sites, the hierarchical porous architecture revealed a highly efficient photocatalytic H_2 evolution efficiency for water splitting under visible light irradiation ($\lambda \geq 420$ nm), which is about nine-fold higher than that of the bulk carbon nitride (BCN). Various characterization approaches were performed to investigate the typical structure formation mechanism and the photoelectric properties of PPCN for the remarkably enhanced hydrogen evolution efficiency.

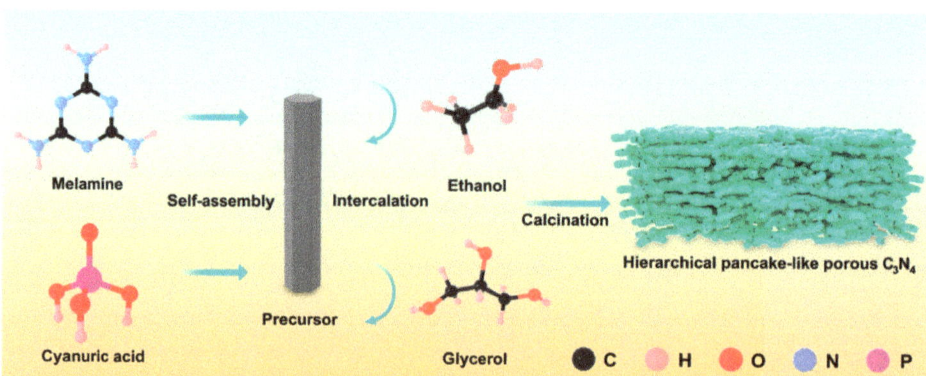

Scheme 1. Schematic illustration of hierarchical pancake-like porous C_3N_4 formation.

2. Results and Discussion

2.1. Morphology and Structure Characterization

The morphologies of BCN and PPCN were investigated via SEM and TEM. It can be seen that BCN appears as a large solid block material (Figure 1a) with seriously agglomerated structure. The light absorption, charge transport, and mass immigration were severely restrained owing to this bulky structure. In contrast, a 3D hierarchical pancake-like porous carbon nitride (Figure 1b–d) exhibited the opposite morphology with an enlarged dimension and abundant pores, which is greatly favorable for the dispersion and diffraction of the light toward the enhanced light absorption capacity. Meanwhile, TEM images (Figure 1e–g) displayed the mesoporous and lattice distance (Figure 1h) of 0.348 nm.

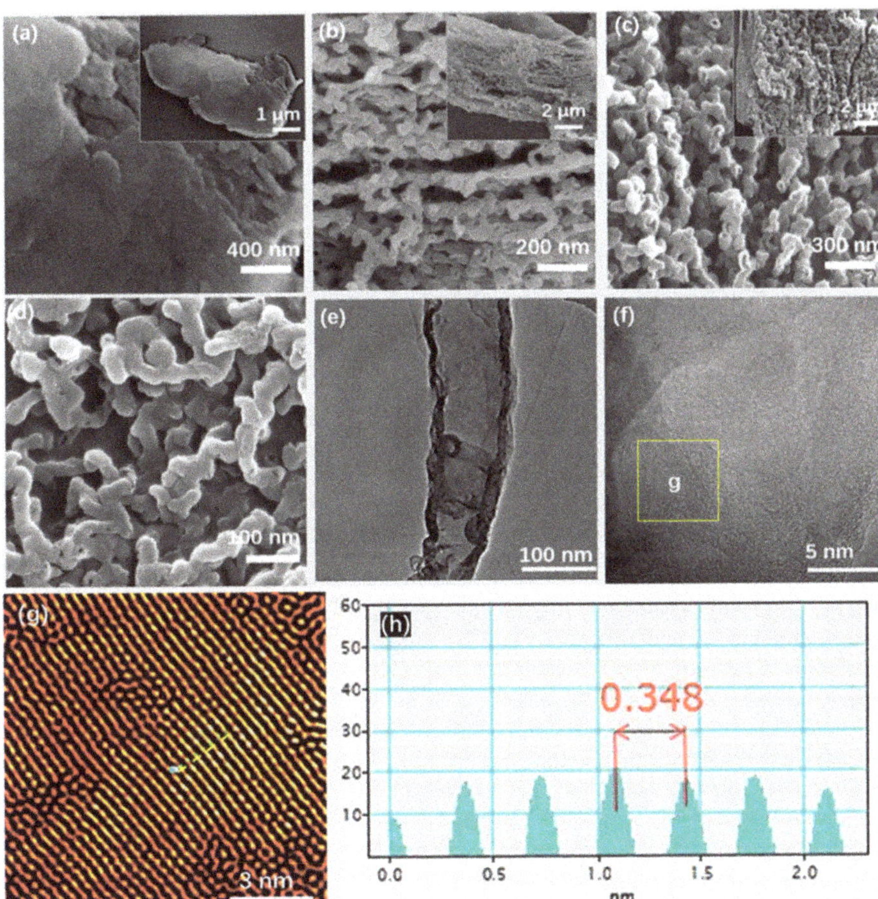

Figure 1. (**a**) SEM image of BCN, (**b–d**) SEM images of PPCN, (**e–g**) TEM images, and (**h**) lattice distance pattern of PPCN.

The structure can be exfoliated into hierarchical porous architecture and generated abundant pores using a suitable hybrid of alcohol and glycerol as the imbedding reagents which supporting abundant gas at the polycondensation process, leading to a hierarchical pancake-like porous architectural carbon nitride formation. N_2 adsorption/desorption isotherms (Figure 2a) was performed to prove this unique structure. The calculated BET specific surface area of PPCN is 79.663 m^2g^{-1}, which is seven-fold higher than that BCN (10.643 m^2g^{-1}). Meanwhile, the peak intensity centered at 37.2 nm in BJH pore distribution curves of PPCN sample (Figure 2b) also exhibited a larger value than BCN implying an abundant nanoporosity.

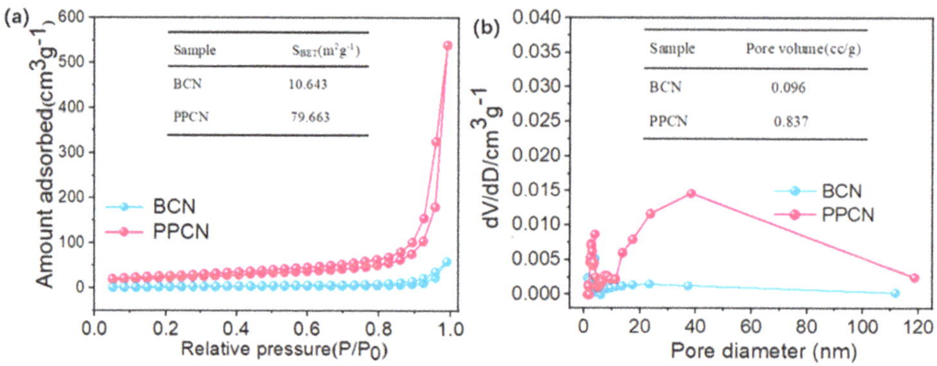

Figure 2. (**a**) N_2 adsorption-desorption isotherm curves and (**b**) BJH pore size distribution curves of BCN and PPCN.

The XRD patterns of PPCN is shown in Figure 3a. Compared to the BCN, PPCN presented a broader and weaker diffraction peak located at 27.2° (002), which is originated from the stacking reflection of interlayer toward a reduced dimension structure [53]. Meanwhile, the peak at 12.9°, ascribed to the repeats in the plane of tri-s-triazine unit could hardly been observed owing to the abundant pores in the interface. Fouriertransform infrared (FT-IR) spectroscopy (Figure 3b) demonstrates that the PPCN possessing a similar chemical structure with BCN. The 810 cm^{-1} peaks are characteristic signal of triazine unit, and the peaks ranges from 1200 to 1600 cm^{-1} is corresponding to the C-N heterocycles. In addition, the broad band peaks located at 3000–3600 cm^{-1} were contributed to-NH stretching vibrations [54]. It is notably that the terminal-NH absorption peak at 3000–3600 cm^{-1} of PPCN vibrate more strongly compared to that of BCN, which can be attributed to the difference in the crystal structural between BCN and PPCN.

X-ray photoelectron spectroscopy (XPS) was further performed to investigate the composition and chemical states of BCN and PPCN. As can be seen in Figure 3c, the presence peaks of C, N, and O are found both in the PPCN and BCN. It should be noted that the higher intensity of O 1s signal of PPCN is easy for the absorption of water and CO_2 due to the hierarchical porous architecture structure. The corresponding high-resolution spectra of C 1s, N 1s, O1s are shown in Figure 3d–f, respectively. The C 1s XPS spectrum revealed three peaks, which can be ascribed to the adventitious carbon species (284.4 eV) [55], C–NH_X species on the edges of aromatic rings (286.03 eV) [56], and the sp^2-hybridized C atom in N==C(–N)$_2$ (288.0 eV) [57], which is consistent with that of BCN. The intensity of the sp^2-bonded carbon of PPCN was increased compared to the BCN may be attributed to the loss of lattice nitrogen. The N 1s XPS signals can be deconvolved into three peaks at398.46, 399.75, and 400.93 eV, which originated from the sp^2-hybridized N attached to C–N=C groups, bridging N in N–(C)$_3$ groups, and the amine functional groups (C–N–H), respectively [58,59]. Additionally, the O 1s spectra can be divided into two peaks located at 532.68 and 530.71 eV, which belongs to the adsorbed water and CO_2 [60].

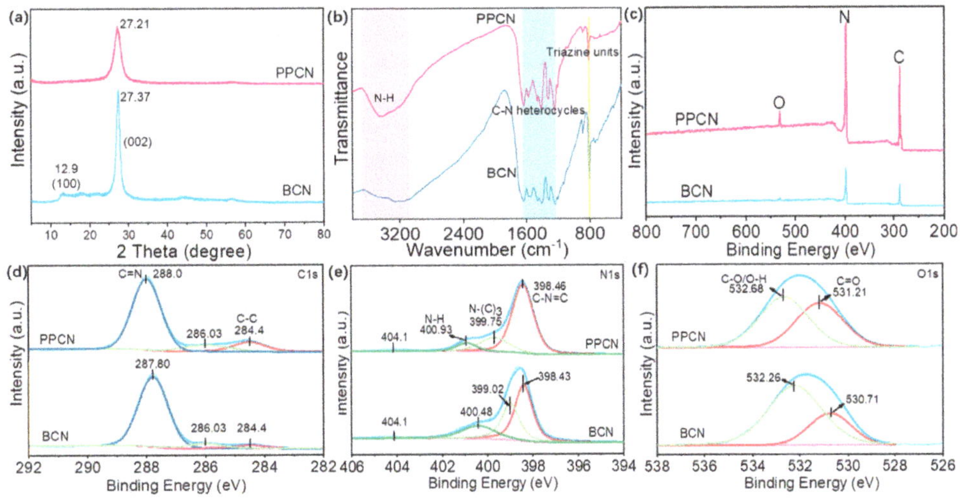

Figure 3. (**a**) XRD and (**b**) FT-IR spectra of BCN and PPCN, Survey and high-resolution XPS spectra of BCN and PPCN, (**c**) survey, (**d**) C 1s, (**e**) N 1s, and (**f**) O 1s of BCN and PPCN.

2.2. Photoelectric Property

UV–VIS diffuse reflectance (DRS) was used to assess the optical absorption properties of BCN and PPCN. Figure 4a indicates that BCN has an absorption edge at 460 nm. Compared with BCN, the PPCN edge shows apparent blue shift, indicating the increase in band gap of semiconductors caused by the quantum confinement effect. The bandgaps for PPCN and BCN were 2.97 and 2.80 eV according to the Kubelka–Munk function [61] (Figure 4b). Consequently, the enlarged bandgap of PPCN would endow the photo-induced electrons and holes with relatively stronger redox abilities toward photocatalytic reactions. The energy band structure of materials can be determined by the conduction band or valence band together with the band gap. In this article, we use the valence band to estimate the conduction band position [62]. The valence band (VB) in Figure 4c of BCN and PPCN was obtained by the XPS valence band spectra. From the band gap date in Figure 4b, the valence band potentials of PPCN and BCN were 1.98 and 2.30 eV vs. NHE, respectively. Finally, the energy band structure was depicted in Figure 4d. Obviously, the more negative conduction band levels of PPCN would contribute to a more powerful e^- reduction ability owing to the more negative CB.

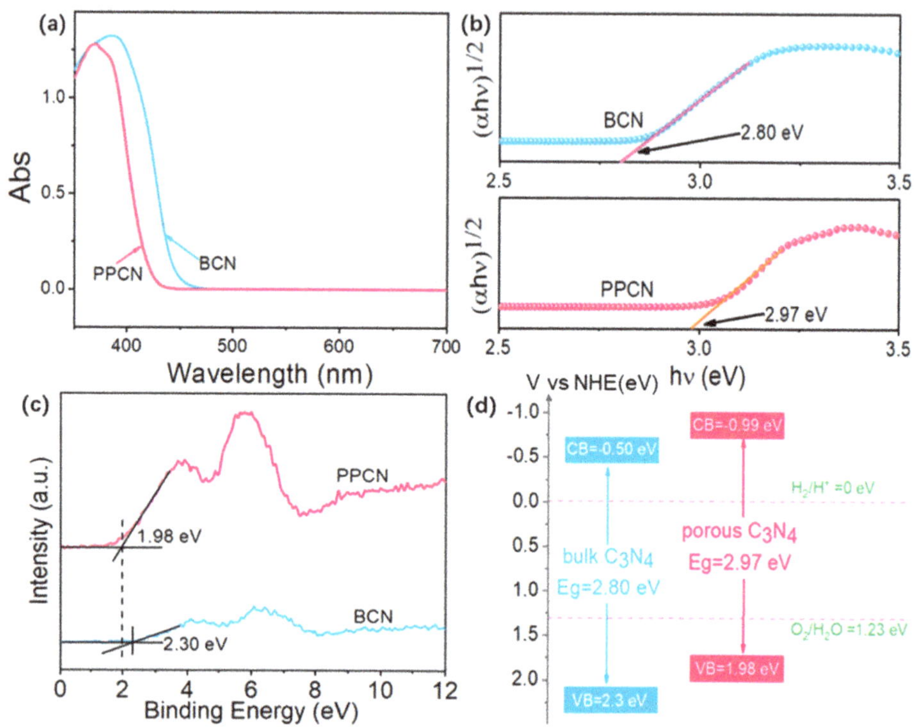

Figure 4. (**a**) UV–VIS DRS spectra and (**b**) Kubelka-Munk transformed function of BCN and PPCN, (**c**) valence band XPS spectra, (**d**) energy band structure of BCN and PPCN.

The photoluminescence tests were further performed with a 325 nm excitation wavelength to assess the separation efficiency of photo-induced electron-hole pairs. As shown in Figure 5a, PPCN presented a lower emission intensity compared with BCN, which indicated of reduced recombination of photo-induced electron-hole pairs in PPCN. It is apparent from this image that the absorption peak of PPCN shifted to a lower value resulted from the quantum size confinement, which is consistent with the DRS.

The time-resolved PL decay spectra were also performed to evaluate the photo-physical characteristics of photogenerated charge carriers of BCN and PPCN. As shown in Figure 5b, the PPCN exhibited a slower exponential decay and fluorescence than that of BCN. τ_1 is originated from the free excitons recombination in semiconductor and τ_2 is ascribed to the non-illuminated recombination of charge carriers in the surface defect states [63]. The short-lived and long-lived lifetime of charge carriers in PPCN (1.743 and 7.595 ns) are both higher than that BCN (1.211 and 5.733 ns). Mostly, the calculated average lifetime in PPCN was 5.089 ns which substantially exceeds 3.696 ns in BCN. The prolonged carriers lifetime and larger percentage of long lifetime charge carriers in PPCN are undoubtedly favorable for promoting the probability of charge carries to involve in photocatalytic reactions. The photocurrent response can also prove this similar conclusion [64]. As depicted in Figure 5c, the PPCN exhibited increased photocurrent intensity than BCN during six light on–off recycles, demonstrating a more efficient separation and transfer of photo-induced electrons. Furthermore, PPCN exhibited a smaller charge transfer resistance (Figure 5d) than that of BCN. Therefore, the hierarchical pancake-like porous architectural structure of PPCN is benefit to the highly efficient separation and migration of photo-induced electron-hole pairs.

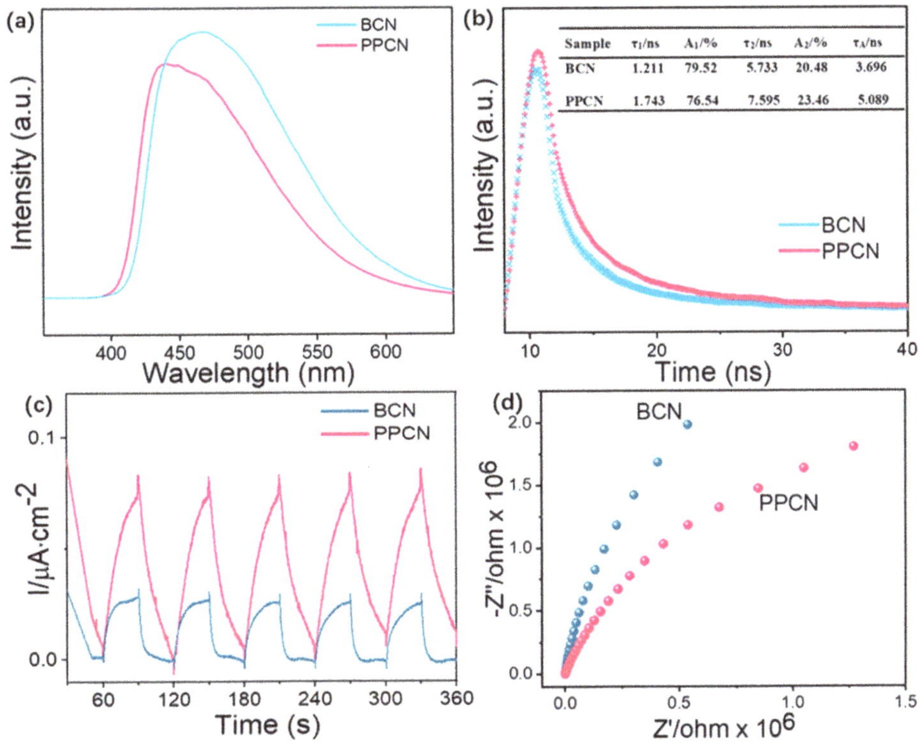

Figure 5. (**a**) Fluorescence emission spectra and (**b**) time-resolved fluorescence excited by incident light of 365 nm of BCN and PPCN, (**c**) photocurrent, and (**d**) EIS spectra of BCN and PPCN.

2.3. Photocatalytic Performance

Though photocatalyst based on carbon nitride without precious Pt is a trend in recent years [65–67], platinum plays a remarkable role in photocatalytic hydrogen production due to the low overpotential and the optimal Gibbs free energy for proton [68,69]. Figure 6a presented the HER of PPCN with 3%Pt loading under visible light ($\lambda \geq 420$ nm) using 15 vol% TEOA as electron donor. As shown in Figure 6a, 3%Pt@PPCN had an HER of 430 μmol g^{-1} h^{-1}, which was nine-fold higher than that of 3%Pt@BCN (45 μmol g^{-1} h^{-1}). The apparent quantum yield (AQE) for the Pt@PPCN reached as high as 19%at 420 nm (see the calculation in Supplementary Materials). Recycling production H_2 evolution test of 3%Pt@PPCN was further performed to assess the reusability of photocatalyst. As can be seen in Figure 6b, the PPCN maintained an excellent HER stability under visible light irradiation, demonstrating the excellent structure stability of PPCN.

Overall, a rational photocatalytic HER mechanism is illustrated in Scheme 2 based on the above results and discussions. Under visible light illumination, the electrons in the valence band (VB) of PPCN are activated to the conduction band (CB) and then are trapped by the Pt for proton reduction. The holes in the VB (+1.98 eV) of PPCN oxidized the sacrificial regent TEOA and converted H_2O into oxidation products (such as H_2O_2). The CB level of PPCN (−0.99 eV) is more negative than that of BCN (0.50 eV), which greatly boosted the reduction capacity of the PPCN. As a consequence, benefited from the 3D hierarchical pancake-like porous architecture, the utilization capacity of light absorption of PPCB would be forcefully enhanced by the scattering and diffraction in the pores and vertical plane, which significantly effective to boost the light-harvesting. More than that, the enlarged specific surface area offered more reactive sites and reduced the recombination rate of photogenerated h^+–e^-pairs, boosting a highly efficient photocatalytic H_2 evolution reaction.

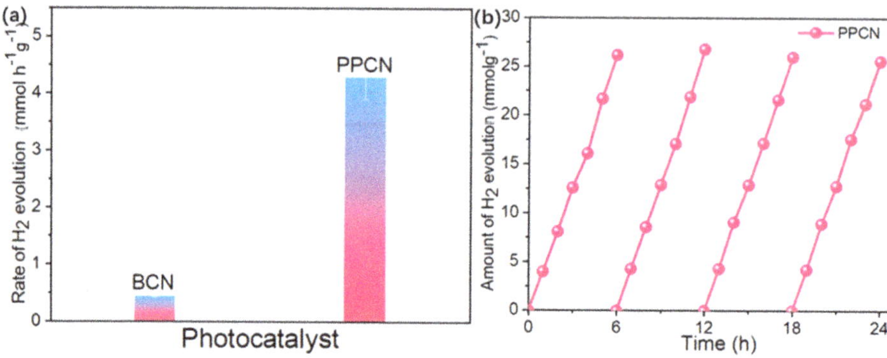

Figure 6. (**a**) H_2 production of the BCN and PPCN with 3%Pt loading, (**b**) cycling H_2 production of 3 wt % Pt@PPCN, Reaction conditions: (**a**,**b**) purified water photocatalytic H_2 production with TEOA as sacrificial under visible light (λ ≥ 420 nm), every 6 h as a period by recovering the catalyst and started with fresh purified water.

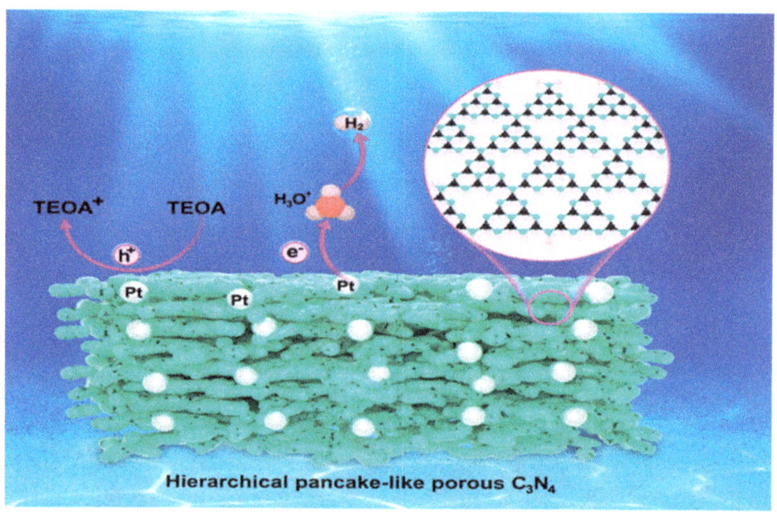

Scheme 2. Proposed photocatalysis charge transfer mechanism of Hierarchical pancake-like porous C_3N_4 in TEOA solution.

3. Experimental

3.1. Reagents

Melamine was purchased from Tianjin Damao Chemical Regent Co., Ltd. (Tianjin, China). Phosphorous acid, glycerol, and ethanol were all supplied by Xilong Scientific Co., Ltd (Jiangsu, China). All reagents used in this study were at least of analytic grade without further purification. DI water was used in the whole experiment.

3.2. Synthesis of Bulk C_3N_4

In the typical procedure, 10.0 g melamine powders were placed into a crucible sealed with a cover followed by calcinating at 550 °C for 4 h in a silica oxide tube furnace with a ramping rate of 2.5 °C/min under N_2 flow (99.999%, 50 mL/min). The as-obtained solid blocks were named bulk C_3N_4.

3.3. Synthesis of Precursor

A total of 1.0 g melamine and 1.2 g phosphorous acid were added to 100 mL deionized water in a magnetic stirrer kept vigorous stirring for 1h at 80 °C. Next, the solution was transferred into Teflon-lined stainless-steel autoclave at 180 °C for 10 h. The mixture was centrifuged, washed with deionized water for several times to obliterate the phosphorus species. Finally, the precursors were obtained after drying at 60 °C.

3.4. Preparation of Hierarchical Pancake-Like Porous Carbon Nitride

Typically, 0.6 g precursor was refluxed with a mixed aqueous solution of 5 mLglycerol and 15 mL ethanol for 3 h in the temperature 90 °C. Then, the powders were washed by ethanol three times and dried at 60 °C. After that, the powders were transferred into a crucible tightly coated with a single layer aluminum foil paper. Finally, the crucible was placed into a muffler and heated to 500 °C for 6 h with a ramping rate of 2 °C/min to obtain the PPCN.

3.5. Characterization

X-ray diffraction (XRD, Shimadzu XRD-6100, Kyoto, Japan) was performed to identify the crystal phase and structural variation of the samples with a Cu–Ka radiation at 40 kV and 30 mA. Scanning electron macroscopy (SEM, Zeiss Sigma, Jena, Germany) and transmission electron microscope (TEM, JEM-2010, JEOL, Tokyo, Janpan) images was recorded to study the morphologies of the samples. Fourier transform infrared (FT-IR) spectra (PerkinElmer Spectrum, Waltham, USA) were recorded to examine the surface functional groups of samples. X-ray photoelectron spectroscopy (XPS) was conducted on thermo ESCALAB 250XI (Axis Ultra DLD Kratos AXIS SUPRA; phi-5000 versaprobe). Nitrogen-adsorption-desorption isotherms (Quantachrome autosorb-IQ2, Quantachrome Instruments, Florida, USA) were performed to measure the specific surface area and pore volume of the samples in a Brunauer-Emmett-Teller (BET) method. UV–VIS diffusion reflectance spectra (DRS) of the samples were recorded on a PerkinElmer Lambda with $BaSO_4$ as a reference. Photoluminescence (PL) spectra were recorded with an excitation wavelength of 325 nm. The time-resolved fluorescence decay spectra were acquired on a Hamamatsu compact fluorescence lifetime spectrometer (C11367, Quantaurus-Tau, Hamamatsu, Japan) with 365 nm excitation wavelength and 470 nm emission wavelength.

3.6. Electrochemical Analysis

The photocurrent and electrochemical impedance spectroscopy (EIS) were investigated with a CHI660C electrochemical workstation equipped with standard three-electrodes in 0.5 M Na_2SO_4 aqueous solution. Among the electrodes, Photocatalyst was deposited on the Fluorine doped Tin Oxide (FTO) electrode and served as the photoanode. Pt wire as the counter electrode and saturated calomel electrode (SCE) as the reference electrode. The preparation approach of photoanode was presented as follows: 50 mg catalyst was dispersed in 100 µL ethanol to obtain a slurry. Next, the slurry was carefully coated onto a 7 mm × 14 mm FTO conductive glass and dried in room temperature. A 500 W Xe lamp quipped with a 420 nm cut-off filter was used as the light irradiation source. The EIS spectrum was recorded using the PPCN as the working electrode at a potential of 0.24 V versus SCE with the frequency ranging from 1 MHz to 0.01 Hz.

3.7. Photocatalytic Hydrogen Evolution

Photocatalytic hydrogen production performance was performed via a sealed reactor and a recirculated cooling water system. Typically, 25 mg of photocatalyst powder was dissolved in 80 mL aqueous solution (68 mL 3 wt % H_2PtCl_6 solution and 12 mL triethanolamine). After that, the reactor was degassed and illuminated 2 h under a full arc light 120 mW·cm^{-2} (λ > 300 nm) with a 420 nm wavelength cut off. Finally, the generated hydrogen was monitored by GC.

4. Conclusions

In summary, a 3D hierarchical pancake-like porous carbon nitride with remarkable photocatalytic H_2 production efficiency was synthesized by a facile bottom-up strategy. 3 wt % @Pt/PPCN showed 430 μmol g^{-1} h^{-1}for photocatalytic H_2 production under visible light ($\lambda \geq$ 420 nm), which was nine times higher than that of 3 wt % @Pt/BCN (45 μmol g^{-1} h^{-1}). The enlarged specific surface area, prolonged charge carriers lifetime, boosted light-harvesting and utilization capacity, and improved charge separation and transfer efficiency are all contributed to the optimized photocatalytic capacity. In addition, this research provides a simple and environmental-friendly approach to design highly-efficient polymeric carbon nitride catalysts for potential applications in solar energy-driven photocatalytic water splitting.

Supplementary Materials: The following are available online at http://www.mdpi.com/2073-4344/10/1/77/s1.

Author Contributions: Conceptualization, X.Q., L.Q. and S.D.; validation, X.Q., L.Q. and S.D.; formal analysis, X.Q.; investigation, X.Q.; resources, X.Q.; data curation, X.Q., M.M. and Y.H.; writing—original draft preparation, X.Q.; writing—review and editing, X.Q.; visualization, X.Q.; supervision, L.Q. and S.D.; project administration, L.Q. and S.D.; funding acquisition, L.Q. and S.D. All authors have read and agreed to the published version of the manuscript.

Funding: The authors gratefully acknowledge the support from the financial support of National Natural Science Foundation of China (no.41763020), the Natural Science Foundation of Jiangxi Province (No.20171BAB206008), the Foundation of Jiangxi Educational Commission (no. GJJ180596, GJJ180603), and the Doctoral Scientific Research Starting Foundation of Jiangxi Science and Technology Normal University (no. 2017BSQD003).

Conflicts of Interest: The authors declare no conflict of interest.

References

1. Asif, M.; Muneer, T. Energy supply, its demand and security issues for developed and emerging economies. *Renew. Sustain. Energy Rev.* **2007**, *11*, 1388–1413. [CrossRef]
2. Wang, X.; Blechert, S.; Antonietti, M. Polymeric Graphitic Carbon Nitride for Heterogeneous Photocatalysis. *ACS Catal.* **2012**, *2*, 1596–1606. [CrossRef]
3. Zhang, C.; Zeng, G.; Huang, D.; Lai, C.; Chen, M.; Cheng, M.; Tang, W.; Tang, L.; Dong, H.; Huang, B.; et al. Biochar for environmental management: Mitigating greenhouse gas emissions, contaminant treatment, and potential negative impacts. *Chem. Eng. J.* **2019**, *373*, 902–922. [CrossRef]
4. Mahmood, J.; Li, F.; Jung, S.M.; Okyay, M.S.; Ahmad, I.; Kim, S.J.; Park, N.; Jeong, H.Y.; Baek, J.B. An efficient and pH-universal ruthenium-based catalyst for the hydrogen evolution reaction. *Nat. Nanotechnol.* **2017**, *12*, 441–446. [CrossRef]
5. Liu, J.; Liu, Y.; Liu, N.; Han, Y.; Zhang, X.; Huang, H.; Lifshitz, Y.; Lee, S.-T.; Zhong, J.; Kang, Z. Metal-free efficient photocatalyst for stable visible water splitting via a two-electron pathway. *Science* **2015**, *347*, 970–974. [CrossRef] [PubMed]
6. Zhang, C.; Lai, C.; Zeng, G.; Huang, D.; Yang, C.; Wang, Y.; Zhou, Y.; Cheng, M. Efficacy of carbonaceous nanocomposites for sorbing ionizable antibiotic sulfamethazine from aqueous solution. *Water Res.* **2016**, *95*, 103–112. [CrossRef]
7. Fujishima, A.; Honda, K. Electrochemical Photolysis of Water at a Semiconductor Electrode. *Nature* **1972**, *238*, 37–38. [CrossRef]
8. Xiong, B.; Chen, L.; Shi, J. Anion-Containing Noble-Metal-Free Bifunctional Electrocatalysts for Overall Water Splitting. *ACS Catal.* **2018**, *8*, 3688–3707. [CrossRef]

9. Wu, J.; Li, N.; Fang, H.-B.; Li, X.; Zheng, Y.-Z.; Tao, X. Nitrogen vacancies modified graphitic carbon nitride: Scalable and one-step fabrication with efficient visible-light-driven hydrogen evolution. *Chem. Eng. J.* **2019**, *358*, 20–29. [CrossRef]
10. Rahman, M.Z.; Davey, K.; Qiao, S.Z. Carbon, nitrogen and phosphorus containing metal-free photocatalysts for hydrogen production: Progress and challenges. *J. Mater. Chem. A* **2018**, *6*, 1305–1322. [CrossRef]
11. Rahman, M.Z.; Davey, K.; Mullins, C.B. Tuning the intrinsic properties of carbon nitride for high Quantum yield photocatalytic hydrogen production. *Adv. Sci.* **2018**, *5*, 1800820. [CrossRef] [PubMed]
12. Rahman, M.Z.; Mullins, C.B. Understanding charge transport in carbon nitride for enhanced photocatalytic solar fuel production. *Acc. Chem. Res.* **2019**, *52*, 248–257. [CrossRef] [PubMed]
13. Yang, J.; Wang, D.; Han, H.; Li, C. Roles of cocatalysts in photocatalysis and photoelectrocatalysis. *Acc. Chem. Res.* **2013**, *46*, 1900–1909. [CrossRef] [PubMed]
14. Walter, M.G.; Warren, E.L.; McKone, J.R.; Boettcher, S.W.; Mi, Q.; Santori, E.A.; Lewis, N.S. Solar water splitting cells. *Chem. Rev.* **2010**, *110*, 6446–6473. [CrossRef] [PubMed]
15. Tu, W.; Zhou, Y.; Zou, Z. Versatile graphene-promoting photocatalytic performance of semiconductors: Basic principles, synthesis, solar energy conversion, and environmental applications. *Adv. Funct. Mater.* **2013**, *23*, 4996–5008. [CrossRef]
16. Kapilashrami, M.; Zhang, Y.; Liu, Y.S.; Hagfeldt, A.; Guo, J. Probing the optical property and electronic structure of TiO_2 nanomaterials for renewable energy applications. *Chem. Rev.* **2014**, *114*, 9662–9707. [CrossRef]
17. Li, X.; Wang, T.; Duan, P.; Baldini, M.; Huang, H.T.; Chen, B.; Juhl, S.J.; Koeplinger, D.; Crespi, V.H.; Schmidt-Rohr, K.; et al. Carbon nitride nanothread crystals derived from pyridine. *J. Am. Chem. Soc.* **2018**, *140*, 4969–4972. [CrossRef]
18. Li, J.; Zhang, Z.; Cui, W.; Wang, H.; Cen, W.; Johnson, G.; Jiang, G.; Zhang, S.; Dong, F. The spatially oriented charge flow and photocatalysis mechanism on internal van der waals heterostructures enhanced g-C_3N_4. *ACS Catal.* **2018**, *8*, 8376–8385. [CrossRef]
19. Huang, D.; Luo, H.; Zhang, C.; Zeng, G.; Lai, C.; Cheng, M.; Wang, R.; Deng, R.; Xue, W.; Gong, X.; et al. Nonnegligible role of biomass types and its compositions on the formation of persistent free radicals in biochar: Insight into the influences on Fenton-like process. *Chem. Eng. J.* **2019**, *361*, 353–363. [CrossRef]
20. Han, C.; Meng, P.; Waclawik, E.R.; Zhang, C.; Li, X.H.; Yang, H.; Antonietti, M.; Xu, J. Palladium/graphitic carbon nitride (g-C_3N_4) stabilized emulsion microreactor as a store for hydrogen from ammonia borane for use in alkene hydrogenation. *Angew. Chem. Int. Ed.* **2018**, *57*, 14857–14861. [CrossRef]
21. Cao, S.; Yu, J. g-C3N4-Based Photocatalysts for hydrogen generation. *J. Phys. Chem. Lett.* **2014**, *5*, 2101–2107. [CrossRef] [PubMed]
22. Cao, S.; Li, H.; Tong, T.; Chen, H.-C.; Yu, A.; Yu, J.; Chen, H.M. Single-atom engineering of directional charge transfer channels and active sites for photocatalytic hydrogen evolution. *Adv. Funct. Mater.* **2018**, *28*. [CrossRef]
23. Zhang, J.; Chen, X.; Takanabe, K.; Maeda, K.; Domen, K.; Epping, J.D.; Fu, X.; Antonietti, M.; Wang, X. Synthesis of a carbon nitride structure for visible-light catalysis by copolymerization. *Angew. Chem. Int. Ed.* **2010**, *49*, 441–444. [CrossRef] [PubMed]
24. Xiao, Y.; Tian, G.; Li, W.; Xie, Y.; Jiang, B.; Tian, C.; Zhao, D.; Fu, H. Molecule self-assembly synthesis of porous few-layer carbon nitride for highly efficient photoredox catalysis. *J. Am. Chem. Soc.* **2019**, *141*, 2508–2515. [CrossRef] [PubMed]
25. Wen, X.; Sun, N.; Tan, Y.; Wang, W.; Yan, C.; Wang, H. One-step synthesis of petals-like graphitic carbon nitride nanosheets with triazole defects for highly improved photocatalytic hydrogen production. *Int. J. Hydrogen Energy* **2019**, *44*, 2675–2684. [CrossRef]
26. Shu, Z.; Wang, Y.; Wang, W.; Zhou, J.; Li, T.; Liu, X.; Tan, Y.; Zhao, Z. A green one-pot approach for mesoporous g-C_3N_4 nanosheets with in situ sodium doping for enhanced photocatalytic hydrogen evolution. *Int. J. Hydrogen Energy* **2019**, *44*, 748–756. [CrossRef]
27. Liu, Q.; Shen, J.; Yu, X.; Yang, X.; Liu, W.; Yang, J.; Tang, H.; Xu, H.; Li, H.; Li, Y.; et al. Unveiling the origin of boosted photocatalytic hydrogen evolution in simultaneously (S, P, O)-codoped and exfoliated ultrathin g-C_3N_4 nanosheets. *Appl. Catal. B Environ.* **2019**, *248*, 84–94. [CrossRef]
28. Kessler, F.K.; Zheng, Y.; Schwarz, D.; Merschjann, C.; Schnick, W.; Wang, X.; Bojdys, M.J. Functional carbon nitride materials-design strategies for electrochemical devices. *Nat. Rev. Mater.* **2017**, *2*, 17030. [CrossRef]

29. Wang, Y.; Zhang, J.; Wang, X.; Antonietti, M.; Li, H. Boron- and fluorine-containing mesoporous carbon nitride polymers: Metal-free catalysts for cyclohexane oxidation. *Angew. Chem. Int. Ed.* **2010**, *49*, 3356–3359. [CrossRef]
30. Cao, S.; Fan, B.; Feng, Y.; Chen, H.; Jiang, F.; Wang, X. Sulfur-doped g-C_3N_4 nanosheets with carbon vacancies: General synthesis and improved activity for simulated solar-light photocatalytic nitrogen fixation. *Chem. Eng. J.* **2018**, *353*, 147–156. [CrossRef]
31. Yang, C.; Wang, Z.; Lin, T.; Yin, H.; Lu, X.; Wan, D.; Xu, T.; Zheng, C.; Lin, J.; Huang, F.; et al. Core-shell nanostructured black rutile titania as excellent catalyst for hydrogen production enhanced by sulfur doping. *J. Am. Chem. Soc.* **2013**, *135*, 17831–17838. [CrossRef] [PubMed]
32. Asahi, R.; Morikawa, T.; Ohwaki, T.; Aoki, K.; Taga, Y. Visible-light photocatalysis in nitrogen-doped titanium oxides. *Science* **2001**, *293*, 269–271. [CrossRef] [PubMed]
33. Zhou, G.; Shan, Y.; Hu, Y.; Xu, X.; Long, L.; Zhang, J.; Dai, J.; Guo, J.; Shen, J.; Li, S.; et al. Half-metallic carbon nitride nanosheets with micro grid mode resonance structure for efficient photocatalytic hydrogen evolution. *Nat. Commun.* **2018**, *9*, 3366. [CrossRef] [PubMed]
34. Tong, Z.; Yang, D.; Li, Z.; Nan, Y.; Ding, F.; Shen, Y.; Jiang, Z. Thylakoid-inspired multishell g-C_3N_4 nanocapsules with enhanced visible-Light harvesting and electron transfer properties for high-Efficiency photocatalysis. *ACS Nano* **2017**, *11*, 1103–1112. [CrossRef] [PubMed]
35. Youngblood, W.J.; Lee, S.-H.A.; Kobayashi, Y.; Hernandez-Pagan, E.A.; Hoertz, P.G.; Moore, T.A.; Moore, A.L.; Gust, D.; Mallouk, T.E. Photoassisted overall water splitting in a visible light-absorbing dye-sensitized photoelectrochemical Cell. *J. Am. Chem. Soc.* **2009**, *131*, 926–927. [CrossRef]
36. Xue, Y.; Lei, Y.; Liu, X.; Li, Y.; Deng, W.; Wang, F.; Min, S. Highly active dye-sensitized photocatalytic H_2 evolution catalyzed by a single-atom Pt cocatalyst anchored onto g-C_3N_4 nanosheets under long-wavelength visible light irradiation. *New J. Chem.* **2018**, *42*, 14083–14086. [CrossRef]
37. Wang, P.; Guan, Z.; Li, Q.; Yang, J. Efficient visible-light-driven photocatalytic hydrogen production from water by using Eosin Y-sensitized novel g-C_3N_4/Pt/GO composites. *J. Mater. Sci.* **2017**, *53*, 774–786. [CrossRef]
38. Zhu, M.; Kim, S.; Mao, L.; Fujitsuka, M.; Zhang, J.; Wang, X.; Majima, T. Metal-Free Photocatalyst for H_2 evolution in visible to near-infrared region: Black phosphorus/graphitic carbon nitride. *J. Am. Chem. Soc.* **2017**, *139*, 13234–13242. [CrossRef]
39. Sun, J.; Guo, Y.; Wang, Y.; Cao, D.; Tian, S.; Xiao, K.; Mao, R.; Zhao, X. H_2O_2 assisted photoelectrocatalytic degradation of diclofenac sodium at g-C_3N_4/$BiVO_4$ photoanode under visible light irradiation. *Chem. Eng. J.* **2018**, *332*, 312–320. [CrossRef]
40. Liu, Y.; Yang, S.; Yin, S.-N.; Feng, L.; Zang, Y.; Xue, H. In situ construction of fibrous AgNPs/g-C_3N_4 aerogel toward light-driven COx-free methanol dehydrogenation at room temperature. *Chem. Eng. J.* **2018**, *334*, 2401–2407. [CrossRef]
41. Bafaqeer, A.; Tahir, M.; Amin, N.A.S. Well-designed ZnV_2O_6/g-C_3N_4 2D/2D nanosheets heterojunction with faster charges separation via pCN as mediator towards enhanced photocatalytic reduction of CO_2 to fuels. *Appl. Catal. B Environ.* **2019**, *242*, 312–326. [CrossRef]
42. Tan, C.; Cao, X.; Wu, X.J.; He, Q.; Yang, J.; Zhang, X.; Chen, J.; Zhao, W.; Han, S.; Nam, G.H.; et al. Recent Advances in Ultrathin Two-Dimensional Nanomaterials. *Chem. Rev.* **2017**, *117*, 6225–6331. [CrossRef]
43. Rahman, M.Z.; Kwong, C.W.; Davey, K.; Qiao, S.Z. 2D phosphorene as a water splitting photocatalyst: Fundamentals to applications. *Energy Environ. Sci.* **2016**, *9*, 709–728. [CrossRef]
44. Wang, J.; Liu, J.; Yang, H.; Chen, Z.; Lin, J.; Shen, Z.X. Active sites-enriched hierarchical MoS_2 nanotubes: Highly active and stable architecture for boosting hydrogen evolution and lithium storage. *J. Mater. Chem. A* **2016**, *4*, 7565–7572. [CrossRef]
45. Tian, N.; Zhang, Y.; Li, X.; Xiao, K.; Du, X.; Dong, F.; Waterhouse, G.I.N.; Zhang, T.; Huang, H. Precursor-reforming protocol to 3D mesoporous g-C_3N_4 established by ultrathin self-doped nanosheets for superior hydrogen evolution. *Nano Energy* **2017**, *38*, 72–81. [CrossRef]
46. Han, Q.; Wang, B.; Gao, J.; Cheng, Z.; Zhao, Y.; Zhang, Z.; Qu, L. Atomically thin mesoporous nanomesh of graphitic C_3N_4 for high-efficiency photocatalytic hydrogen evolution. *ACS Nano* **2016**, *10*, 2745–2751. [CrossRef]
47. Jiang, Y.; Sun, Z.; Tang, C.; Zhou, Y.; Zeng, L.; Huang, L. Enhancement of photocatalytic hydrogen evolution activity of porous oxygen doped g-C_3N_4 with nitrogen defects induced by changing electron transition. *Appl. Catal. B Environ.* **2019**, *240*, 30–38. [CrossRef]

48. Shalom, M.; Inal, S.; Fettkenhauer, C.; Neher, D.; Antonietti, M. Improving carbon nitride photocatalysis by supramolecular preorganization of monomers. *J. Am. Chem. Soc.* **2013**, *135*, 7118–7121. [CrossRef]
49. Gu, Q.; Gao, Z.; Xue, C. Self-sensitized carbon nitride microspheres for long-lasting visible-light-driven hydrogen generation. *Small* **2016**, *12*, 3543–3549. [CrossRef]
50. Kang, Y.; Yang, Y.; Yin, L.C.; Kang, X.; Liu, G.; Cheng, H.M. An amorphous carbon nitride photocatalyst with greatly extended visible-light-responsive range for photocatalytic hydrogen generation. *Adv. Mater.* **2015**, *27*, 4572–4577. [CrossRef]
51. Rahman, M.Z.; Tapping, P.C.; Kee, T.W.; Smernik, R.; Spooner, N.; Moffatt, J.; Tang, Y.; Davey, K.; Qiao, S.-Z. A benchmark quantum yield for water photoreduction on amorphous carbon nitride. *Adv. Funct. Mater.* **2017**, *27*, 1702384. [CrossRef]
52. Guo, S.; Deng, Z.; Li, M.; Jiang, B.; Tian, C.; Pan, Q.; Fu, H. Phosphorus-doped carbon nitride tubes with a layered micro-nanostructure for enhanced visible-light photocatalytic hydrogen evolution. *Angew. Chem. Int. Ed.* **2016**, *55*, 1830–1834. [CrossRef]
53. Wang, X.; Maeda, K.; Thomas, A.; Takanabe, K.; Xin, G.; Carlsson, J.M.; Domen, K.; Antonietti, M. A metal-free polymeric photocatalyst for hydrogen production from water under visible light. *Nat. Mater.* **2009**, *8*, 76–80. [CrossRef]
54. Niu, P.; Zhang, L.; Liu, G.; Cheng, H.-M. Graphene-like carbon nitride nanosheets for improved photocatalytic activities. *Adv. Funct. Mater.* **2012**, *22*, 4763–4770. [CrossRef]
55. Shinde, S.S.; Lee, C.H.; Yu, J.Y.; Kim, D.H.; Lee, S.U.; Lee, J.H. Hierarchically designed 3D holey C_2N aerogels as bifunctional oxygen electrodes for flexible and rechargeable Zn-Air batteries. *ACS Nano* **2018**, *12*, 596–608. [CrossRef]
56. Wulan, B.-R.; Yi, S.-S.; Li, S.-J.; Duan, Y.-X.; Yan, J.-M.; Jiang, Q. Amorphous nickel pyrophosphate modified graphitic carbon nitride: An efficient photocatalyst for hydrogen generation from water splitting. *Appl. Catal. B Environ.* **2018**, *231*, 43–50. [CrossRef]
57. Shinde, S.S.; Sami, A.; Lee, J.-H. Sulfur mediated graphitic carbon nitride/S-Se-graphene as a metal-free hybrid photocatalyst for pollutant degradation and water splitting. *Carbon* **2016**, *96*, 929–936. [CrossRef]
58. Shen, R.; Xie, J.; Zhang, H.; Zhang, A.; Chen, X.; Li, X. Enhanced solar fuel H_2 generation over g-C_3N_4 nanosheet photocatalysts by the synergetic effect of noble metal-free Co_2P cocatalyst and the environmental phosphorylation strategy. *ACS Sustain. Chem. Eng.* **2017**, *6*, 816–826. [CrossRef]
59. Shinde, S.S.; Sami, A.; Lee, J.-H. Nitrogen- and phosphorus-doped nanoporous graphene/graphitic carbon nitride hybrids as efficient electrocatalysts for hydrogen evolution. *ChemCatChem* **2015**, *7*, 3873–3880. [CrossRef]
60. Tang, H.; Wang, R.; Zhao, C.; Chen, Z.; Yang, X.; Bukhvalov, D.; Lin, Z.; Liu, Q. Oxamide-modified g-C_3N_4 nanostructures: Tailoring surface topography for high-performance visible light photocatalysis. *Chem. Eng. J.* **2019**, *374*, 1064–1075. [CrossRef]
61. Marci, G.; Garcia-Lopez, E.I.; Pomilla, F.R.; Palmisano, L.; Zaffora, A.; Santamaria, M.; Krivtsov, I.; Ilkaeva, M.; Barbierikova, Z.; Brezova, V. Photoelectrochemical and EPR features of polymeric C_3N_4 and O-modified C_3N_4 employed for selective photocatalytic oxidation of alcohols to aldehydes. *Catal. Today* **2019**, *328*, 21–28. [CrossRef]
62. Rahman, M.; Davey, K.; Qiao, S.Z. Counteracting blueshift optical absorption and maximizing photon harvest in carbon nitride nanosheet photocatalyst. *Small* **2017**, *13*, 1700376. [CrossRef]
63. Zeng, Y.; Liu, X.; Liu, C.; Wang, L.; Xia, Y.; Zhang, S.; Luo, S.; Pei, Y. Scalable one-step production of porous oxygen-doped g-C_3N_4 nanorods with effective electron separation for excellent visible-light photocatalytic activity. *Appl. Catal. B Environ.* **2018**, *224*, 1–9. [CrossRef]
64. Yang, L.; Huang, J.; Shi, L.; Cao, L.; Yu, Q.; Jie, Y.; Fei, J.; Ouyang, H.; Ye, J. A surface modification resultant thermally oxidized porous g-C_3N_4 with enhanced photocatalytic hydrogen production. *Appl. Catal. B Environ.* **2017**, *204*, 335–345. [CrossRef]
65. Rahman, M.; Davey, K. Enabling Pt-free photocatalytic hydrogen evolution on polymeric melon: Role of amorphization for overcoming the limiting factors. *Phys. Rev. Mater.* **2018**, *2*, 125402. [CrossRef]
66. Rahman, M.Z.; Tang, Y.; Kwong, P. Reduced recombination and low-resistive transport of electrons for photo-redox reactions in metal-free hybrid photocatalyst. *Appl. Phys. Lett.* **2018**, *112*, 253902. [CrossRef]

67. Rahman, M.Z.; Moffatt, J.; Spooner, N. Topological carbon nitride: Localized photon absorption and delocalized charge carrier separation at intertwined photocatalyst interfaces. *Mater. Horiz.* **2018**, *5*, 553–559. [CrossRef]
68. Zhu, M.; Osakada, Y.; Kim, S.; Fujitsuka, M.; Majima, T. Black phosphorus: A promising two dimensional visible and near-infrared-activated photocatalyst for hydrogen evolution. *Appl. Catal. B Environ.* **2017**, *217*, 285–292. [CrossRef]
69. Jiang, B.; Sun, Y.; Liao, F.; Shen, W.; Lin, H.; Wang, H.; Shao, M. Rh-Ag-Si ternary composites: Highly active hydrogen evolution electrocatalysts over Pt–Ag–Si. *J. Mater. Chem. A* **2017**, *5*, 1623–1628. [CrossRef]

catalysts

Article

Carbon Nitride-Perovskite Composites: Evaluation and Optimization of Photocatalytic Hydrogen Evolution in Saccharides Aqueous Solution

Andrea Speltini [1,*], **Lidia Romani** [2], **Daniele Dondi** [2], **Lorenzo Malavasi** [2] and **Antonella Profumo** [2,*]

1 Department of Drug Sciences, University of Pavia, via Taramelli 12, 27100 Pavia, Italy
2 Department of Chemistry, University of Pavia, via Taramelli 12, 27100 Pavia, Italy; lidia.romani01@universitadipavia.it (L.R.); daniele.dondi@unipv.it (D.D.); lorenzo.malavasi@unipv.it (L.M.)
* Correspondence: andrea.speltini@unipv.it (A.S.); antonella.profumo@unipv.it (A.P.); Tel.: +39-0382-987349 (A.S.)

Received: 29 September 2020; Accepted: 29 October 2020; Published: 30 October 2020

Abstract: The application of hybrid photocatalysts made of carbon nitride and lead-free perovskites, namely $DMASnBr_3/g\text{-}C_3N_4$ and $PEA_2SnBr_4/g\text{-}C_3N_4$, for the H_2 evolution from saccharides aqueous solution is described. The novel composites were tested and compared in terms of hydrogen evolution rate (HER) under simulated solar light, using Pt as a reference co-catalyst, and glucose as a representative sacrificial biomass. The conditions were optimized to maximize H_2 generation by a design of experiments involving catalyst amount, glucose concentration and Pt loading. For both materials, such parameters affected significantly H_2 photogeneration, with the best performance observed using 0.5 g L^{-1} catalyst, 0.2 M glucose and 0.5 wt% Pt. Under optimized conditions, $DMASnBr_3/g\text{-}C_3N_4$ showed a 5-fold higher HER compared to $PEA_2SnBr_4/g\text{-}C_3N_4$, i.e., 925 μmoles g^{-1} h^{-1} and 190 μmoles g^{-1} h^{-1}, respectively (RSD ≤ 11%, n = 4). The former composite, which affords an HER 15-fold higher in aqueous glucose than in neat water, provided H_2 also with no metal co-catalyst (around 140 μmoles g^{-1} h^{-1}), and it was reusable for at least three photoreactions. Encouraging results were also collected by explorative tests on raw starch solution (around 150 μmoles g^{-1} h^{-1}).

Keywords: hydrogen; catalyst; photocatalysis; biomass; solar light; perovskite; carbon nitride; design of experiments

1. Introduction

The search for new photocatalytic systems working under solar light for hydrogen production is increasingly triggering the interest of the scientific community. In the framework of novel photocatalysts, graphitic carbon nitride ($g\text{-}C_3N_4$) has emerged in the last decade as one of the most promising material to run H_2 photoproduction from water under visible light, due to cost-effective and easy synthesis, chemical stability, narrow band-gap and band potentials suitable to perform relevant redox reactions in aqueous solution [1,2]. Its intrinsic limitations, namely relatively high charge carriers recombination, low surface area and restricted light harvesting in the visible region, have been in part compensated by several strategies. For instance, metal and non-metal doping, structural and morphological modifications, dye-sensitization, and combination with co-catalysts of different nature (e.g., carbon nanotubes, carbon dots, bimetallic deposition) have proved to be rewarding in terms of enhanced photocatalytic activity towards H_2 generation from water also in the presence of sacrificial agents [2,3]. Among these, mainly fine chemicals have been used, as recently reviewed by Nasir et al. [2], while just a few studies were undertaken in aqueous biomass solutions or directly in wastewaters [4–6].

In recent times, the application of metal halide perovskites (MHPs) in the photocatalysis field has aroused great interest in view of the relatively narrow bandgaps compared to traditional metal oxide photocatalysts [7,8]. MHPs possess exclusive optical properties, viz. consistent light absorption in the visible region, tunable band-gap and extended carriers lifetimes, conveniently exploited in photovoltaics and optoelectronics [9].

Recently, coupling MHPs with g-C_3N_4 has become a cutting-edge research field [2,10,11], although limited work has been done so far concerning photocatalysis [2]. Recently, lead-free MHPs showing an exceptional stability in water were synthesized and characterized [12–14], thus opening a new avenue for the preparation and application of innovative photocatalysts. In this context, our group advantageously coupled dimethylammonium and phenylethylammonium-based perovskites—$DMASnBr_3$ and PEA_2SnBr_4, respectively—to g-C_3N_4, providing new micro-sized composites with excellent photocatalytic properties towards H_2 production from triethanolamine (TEOA) aqueous solution; in particular, up to a 20-fold increase of hydrogen evolution rate (HER) was achieved compared to g-C_3N_4 alone, due to a synergistic effect between the two constituents in the composite [12,13], essentially due to improved charge carrier separation. Such an effect results from the positive band-alignment between the two semiconductors, namely the MHP and g-C_3N_4, and the perovskite is selected based on its bandgap in order to exploit this synergistic effect. It was also observed in an explorative test that such catalytic systems are capable of generating H_2 from glucose solution under UV-visible light, with HERs from 30 to 100 times higher relative to pristine g-C_3N_4 [12,13].

In this study, the two best performing materials we previously identified, namely 5% PEA_2SnBr_4/g-C_3N_4 and 33% $DMASnBr_3$/g-C_3N_4, were systematically tested, under simulated solar light, in aqueous glucose through a design of experiments (DoE) with the aim to maximize HER and compare the performance of each photocatalyst under the best conditions. The most performant composite was further investigated for its photo-chemical stability by recycling tests, compared to the well-known Evonik Aeroxide® P25 TiO_2, used without any metal co-catalyst, and in the presence of soluble starch as a low-cost and abundant polysaccharide.

2. Results and Discussion

Based on the preliminary findings obtained with carbon nitride-perovskite composites in TEOA solution [12,13], the two best performing materials, 33% $DMASnBr_3$/g-C_3N_4 and 5% PEA_2SnBr_4/g-C_3N_4, are investigated here for a systematic study of H_2 photoproduction from glucose aqueous solution. Glucose was selected as a biomass-derived sacrificial agent because in the wastewaters from food industry sugars are present at considerable amounts [4,5], and Pt was used as the reference metal co-catalyst because of its excellent properties for water reduction due to the large work function, resulting in a strong Schottky barrier effect [15,16].

2.1. Comparison between $DMASnBr_3$/g-C_3N_4 and PEA_2SnBr_4/g-C_3N_4 after Chemometric Optimization

With the aim of investigating the behavior of the two catalysts and maximizing H_2 evolution, a chemometric approach was chosen to easily individuate the effects of the key operational parameters of heterogeneous photocatalysis, namely the relative amounts of catalyst, glucose and metal [17–20]. Accordingly, a full 2^3 experimental design was setup considering glucose concentration (x_1), catalyst amount (x_2) and Pt loading (x_3). As reported in Table 1, the experimental domain was comprised of two levels (−1 and +1) of each variable.

Table 1. Experimental domain for the 2^3 factorial design.

Variable	Level Codes	
	−1	+1
Glucose concentration (M), x_1	0.025	0.2
Catalyst amount (g L^{-1}), x_2	0.5	2
Pt loading (wt%), x_3	0.5	3

HERs obtained under the different conditions (mean values of duplicate tests), according to the experimental matrix, are reported in Table 2.

Table 2. Mean hydrogen evolution rates (HERs) obtained in the conditions of the experimental plan for both composite photocatalysts.

Exp	Glucose Concentration (M), x_1	Catalyst Amount (g L^{-1}), x_2	Pt Loading (wt%), x_3	$DMASnBr_3/g\text{-}C_3N_4$ HER (μmoles g^{-1} h^{-1})	$PEA_2SnBr_4/g\text{-}C_3N_4$ HER (μmoles g^{-1} h^{-1})
1	0.2	2	3	143	27
2	0.025	2	3	128	14
3	0.2	0.5	3	696	99
4	0.025	0.5	3	341	100
5	0.2	2	0.5	194	147
6	0.025	2	0.5	92	43
7	**0.2**	**0.5**	**0.5**	**925**	**191**
8	0.025	0.5	0.5	606	188

The values were used as the experimental response (y), and they were modeled (Matlab® software) relative to each variable (x_i), in line with the following equation:

$$y = b_0 + b_1x_1 + b_2x_2 + b_3x_3 + b_{12}x_1x_2 + b_{13}x_1x_3 + b_{23}x_2x_3, \quad (1)$$

Figure 1 shows the plot of the coefficients (b_i) of the model, while the response surfaces are gathered in Figure 2.

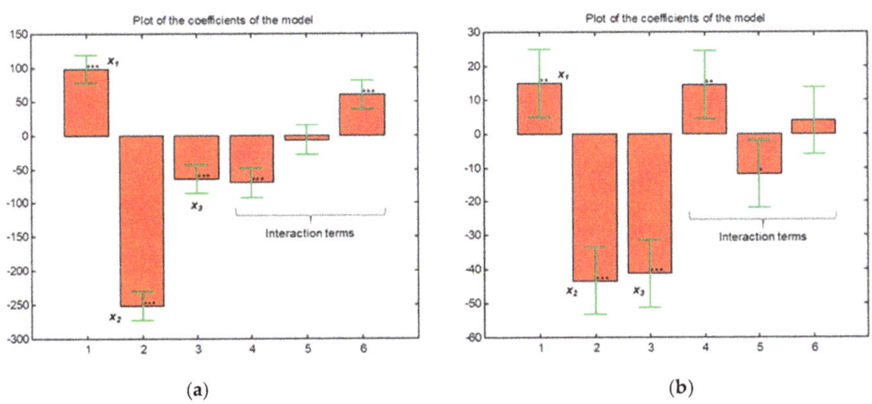

Figure 1. Plots of the coefficients of the model for (**a**) $DMASnBr_3/g\text{-}C_3N_4$ and (**b**) $PEA_2SnBr_4/g\text{-}C_3N_4$. Stars indicate the significance of the coefficients (* $p < 0.05$, ** $p < 0.01$, *** $p < 0.001$), while error bars indicate the confidence intervals ($p = 0.05$).

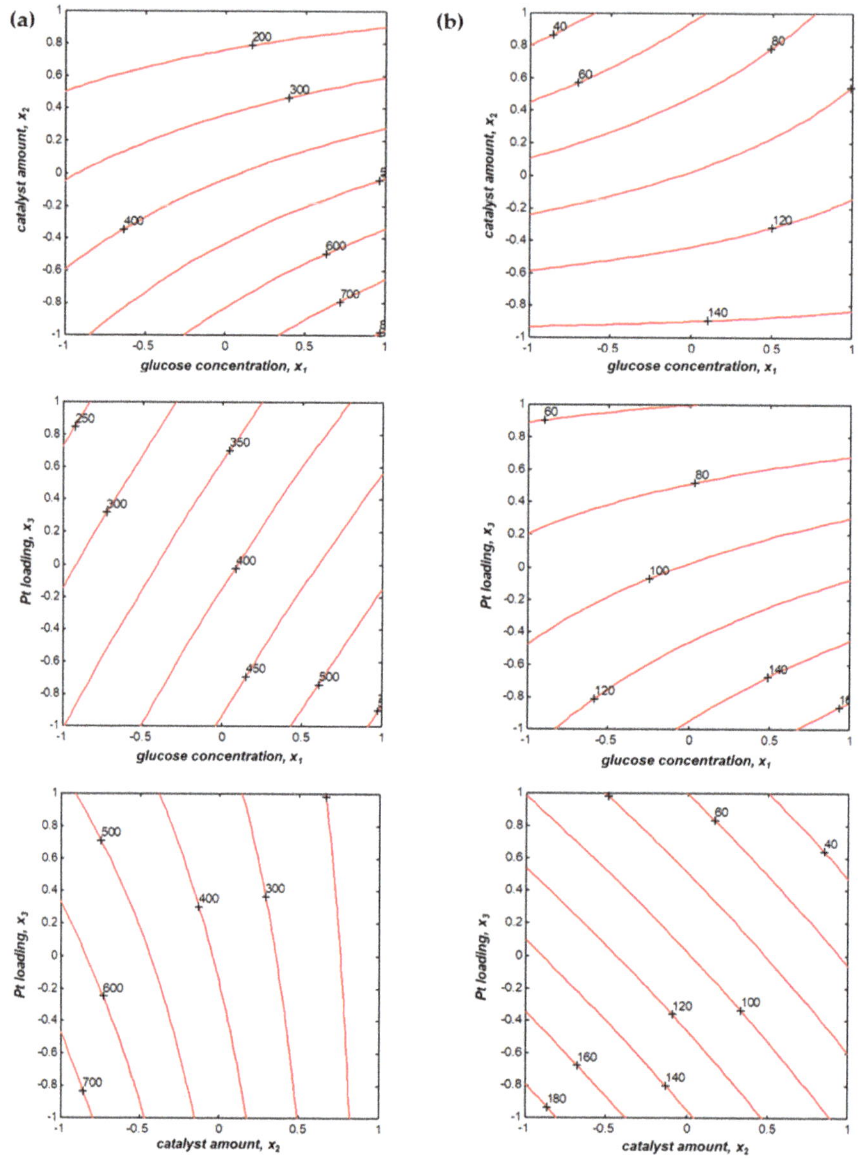

Figure 2. Response surfaces obtained by Matlab® elaboration for (**a**) $DMASnBr_3/g\text{-}C_3N_4$ and (**b**) $PEA_2SnBr_4/g\text{-}C_3N_4$ (for level codes −1 +1 see Table 1); the number close to each curve represents the HER.

The models elaborated on the experimental results are:

$$HER = 390 + 99x_1 - 251x_2 - 64x_3 - 70x_1x_2 - 6x_1x_3 + 60x_2x_3 \quad (2)$$

$$HER = 101 + 15x_1 - 43x_2 - 41x_3 + 14x_1x_2 - 12x_1x_3 + 4x_2x_3 \quad (3)$$

for $DMASnBr_3/g\text{-}C_3N_4$ and $PEA_2SnBr_4/g\text{-}C_3N_4$, respectively.

As apparent from the significance of the coefficients (Figure 1), the picture is similar for both composites, as the three variables affect significantly the HER ($p < 0.001$), although glucose concentration to a minor extent for $PEA_2SnBr_4/g\text{-}C_3N_4$ ($p < 0.01$); in particular, the photoreaction is favored by high glucose concentration and low levels of catalyst in the suspension and metal photodeposited on the catalyst surface. In addition, some interactions are statistically relevant, in particular x_1–x_2 and x_2–x_3 for $DMASnBr_3/g\text{-}C_3N_4$. In line with the response surfaces (see Figure 2), HER increases by keeping the catalyst amount at the low level but with the highest concentration of sacrificial agent, whereas both catalyst and co-catalyst at the lowest level enhance H_2 evolution. For $PEA_2SnBr_4/g\text{-}C_3N_4$, x_1–x_2 and x_1–x_3 are significative ($p < 0.01$ and $p < 0.05$, respectively); accordingly, the response surfaces show that also in this case, HER is maximized with low Pt loading and catalyst amount, and the high level of sacrificial biomass is useful, especially when working with 0.5 wt% Pt.

These findings can be explained considering that at low sacrificial substrate concentration, the mass transfer of glucose from water to the catalyst surface is the bottleneck of photocatalytic reaction, whilst in concentrated solutions, interfacial reactions govern the process, due to the saturation of glucose on the catalyst [17]. Furthermore, excessive Pt loading reduces the catalytic surface available for light absorption; additionally, a high concentration of suspended powder causes scattering of the incident radiation, thus decreasing the overall process [15,18].

Based on these outcomes, the selected conditions were 0.2 M glucose, 0.5 g L^{-1} catalyst, and 0.5 wt% Pt. These are advantageous, involving the use of small amounts of both catalyst and metal co-catalyst, and suggesting that large sample dilution could be avoided when using food industry wastewaters, where the total sugar content is up to tens of grams per liter [5].

The reproducibility was good, with relative standard deviations (RSD) ≤11% and ≤15% observed on four independent experiments for $DMASnBr_3/g\text{-}C_3N_4$ and $PEA_2SnBr_4/g\text{-}C_3N_4$, correspondingly.

For the best performing composite, $DMASnBr_3/g\text{-}C_3N_4$, the optimal conditions were further confirmed by validation of the model, working at the test point ($x_1 = 0$; $x_2 = 0$; $x_3 = 0$). Being the mean HER (407 ± 69 μmoles g^{-1} h^{-1}, $p = 0.05$, $n = 4$) not significantly different from the value predicted by Equation (1), viz. 390 μmoles g^{-1} h^{-1}, it is demonstrated that the model is effective and it can be applied to the whole experimental domain.

The chemometric study provided key information. First, it enabled pointing out the most convenient conditions for H_2 evolution, which resulted in a three-fold increase in HER relative to the first pilot work [13]; at the same time, the results collected clearly highlight the superior photocatalytic activity of $DMASnBr_3/g\text{-}C_3N_4$.

Given the very small difference between the MHPs' band gaps—2.85 and 2.74 eV for $DMASnBr_3$ and PEA_2SnBr_4, respectively [12,13]—and the higher surface area of $PEA_2SnBr_4/g\text{-}C_3N_4$ (5.5 m^2 g^{-1} vs. 3.4 m^2 g^{-1} of $DMASnBr_3/g\text{-}C_3N_4$), it could be speculated that the better performance of the first catalyst derives from the different band alignments of the MHPs' valence band (VB) and conduction band (CB) edges with those of carbon nitride (gathered in Figure 3). This results in a different electron transfer and charge carriers stabilization mechanism.

The synergistic effect between the two constituents of the $DMASnBr_3$-based photocatalyst, which works as a Z-scheme heterojuction, has been explained in terms of favorable band alignment and prolonged charge carrier lifetimes [13]. In the case of PEA_2SnBr_4, this is also true, but it is known that 2D perovskites have higher excitonic binding energies that may slightly reduce the charge carrier dynamics, thus affording a lower photocatalytic efficiency [21].

Another reason rationally stands in the better dispersibility of the DMA composite in aqueous phase due to its lower hydrophobicity. Because of the low surface area of the composites, good mechanical stirring is important; on the other hand, the fast sedimentation of the powders can be an advantage to recover them from water after photocatalysis.

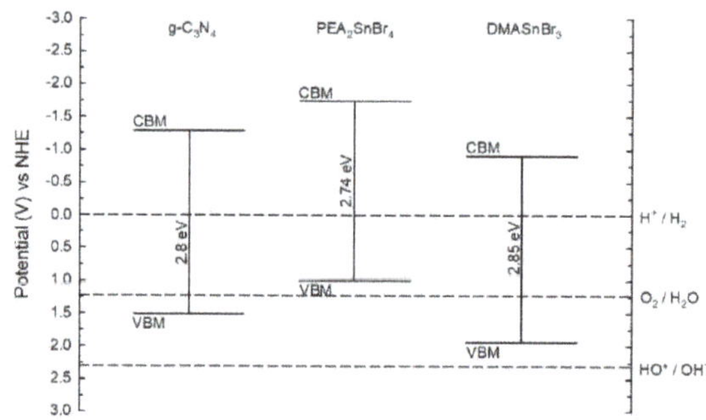

Figure 3. Scheme of the redox potentials (pH 7 vs NHE) of relevant reactions in relation to the band edges positions of as-prepared g-C_3N_4 and metal halide perovskites (MHPs).

2.2. Further Investigation of DMASnBr$_3$/g-C_3N_4

The photocatalytic behavior of the composite containing the DMASnBr$_3$ perovskite was further studied. Table 3 summarizes the HERs collected by key tests aimed at assessing the role of the metal co-catalyst, of the glucose biomass and to evaluate the contribute of "direct water splitting", i.e., H_2 evolved from water in the absence of the sacrificial donor [22].

Table 3. Key experiments for evaluating the photocatalytic system based on 33% DMASnBr$_3$/g-C_3N_4 (simulated solar light).

Sample	HER (μmoles g^{-1} h^{-1}) [1]
water + 0.5 g L^{-1} catalyst	12
water + 0.5 g L^{-1} catalyst + 0.5 wt% Pt	62
0.2 M glucose + 0.5 g L^{-1} catalyst	142
0.2 M glucose + 0.5 g L^{-1} catalyst + 0.5 wt% Pt	**925**
water	n.q.
0.2 M glucose	n.q.

[1] RSDs ≤ 11% (n = 3); n.q.; not quantifiable (<0.008 μmoles h^{-1}).

As is apparent, the model biomass has a major role in sustaining H_2 evolution, which is 15-fold higher relative to the HER observed in neat water (the contribution of "direct water splitting" is <7%). In addition, the metal co-catalyst largely rules the photoreaction to give gas-phase H_2, both in the presence of the sacrificial agent and in pure water. Notice that appreciable HER was gained also with no metal, underlining the potentiality of such photoactive material. This evidence corroborates a photoreaction mechanism typical of the photocatalytic systems for H_2 production from water [1,3,23], wherein:

(1) The catalyst absorbs the radiation generating the characteristic charge separation (holes and electrons);

(2) The sacrificial organic material serves as scavenger of the oxidizing species and, by undergoing gradual oxidation (photoreforming), supplies electrons;

(3) The metal works as an electron collector and is the active site for hydrogen ions reduction while hindering charge carriers' recombination and backward reactions [3].

Regarding the DMASnBr$_3$/g-C_3N_4 composite, valence band holes—not hydroxyl radicals in the solution—are generated in g-C_3N_4 and these trigger biomass oxidation to boost H_2 formation from water, while avoiding a complete aqueous phase reforming of the organic substrate until mineralization [1], which instead occurs in titanium dioxide photocatalysis, especially with Pd co-catalyst [16,22,24]. The negligible H_2 evolution (<0.008 μmoles h^{-1}) observed by irradiation of neat water and 0.2 M glucose, as the control tests (Table 3), further substantiated the proposed mechanism.

The synergistic effect between carbon nitride and perovskite described above and assessed in foregoing work [13] was here confirmed in glucose solution under the best conditions. Indeed, HER was 74 and 40 μmoles g^{-1} h^{-1} for g-C_3N_4 and $DMASnBr_3$ individually tested, respectively, against 925 μmoles g^{-1} h^{-1} of the hybrid catalyst (see Table 3).

The values of apparent quantum yield (AQY), calculated as the percent ratio H_2 moles/incident photons moles [5] and turn over number (TON), calculated as the ratio H_2 moles/Pt moles [16], for $DMASnBr_3$/g-C_3N_4 in the optimized conditions are reported in Table 4.

Table 4. Apparent quantum yield (AQY) and turn over number (TON) values for $DMASnBr_3$/g-C_3N_4 (0.5 g L^{-1} catalyst, 0.5 wt% Pt).

Sample	AQY	TON	HER (μmoles g^{-1} h^{-1})
distilled water	0.1	202	62
aqueous glucose	2.0	3007	925
aqueous starch	0.3	473	146

These results are very interesting and highlight the efficiency of the composite catalyst that, despite the very low surface area (<4 m^2 g^{-1}, 18-times lower than that of the commercial nanometric P25 TiO_2 [25]), promoted HER equal to about one third of the latter (2906 μmoles g^{-1} h^{-1}), under the same conditions. To point out the role of surface area and to provide comparable data, we tested a lower amount of P25 TiO_2 (0.028 g L^{-1}) to have approximately the same catalytic surface area of 0.5 g L^{-1} of the composite. HER was 377 μmoles g^{-1} h^{-1}, 2.5-times lower than that afforded by employing the composite.

Additional trials were devoted to investigate the stability of the catalyst upon subsequent irradiations. As is shown in Figure 4, no loss of efficiency was noticed using the recycled powder in a new sample solution, and only a 25% decrease in H_2 evolution in a third treatment.

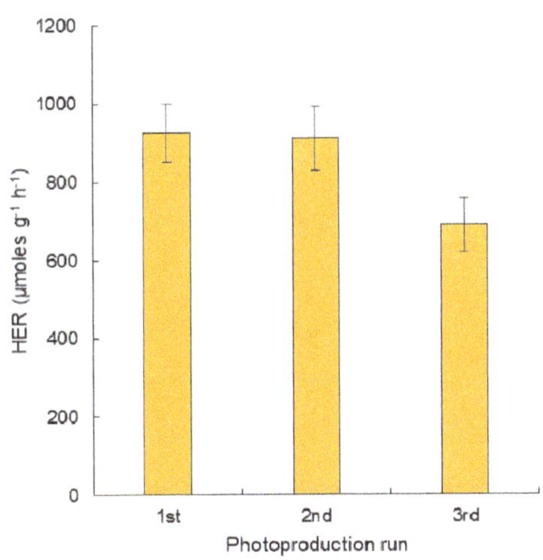

Figure 4. Reusability of $DMASnBr_3$/g-C_3N_4 (0.5 g L^{-1}, 0.5 wt% Pt, 0.2 M aqueous glucose), RSD < 11% (*n* = 3).

The X-ray diffraction (XRD) profiles overlaid in Figure 5 attest the preservation of the pristine structure in the composite employed after the three sequential photoreactions (18 h total irradiation time) in aqueous glucose.

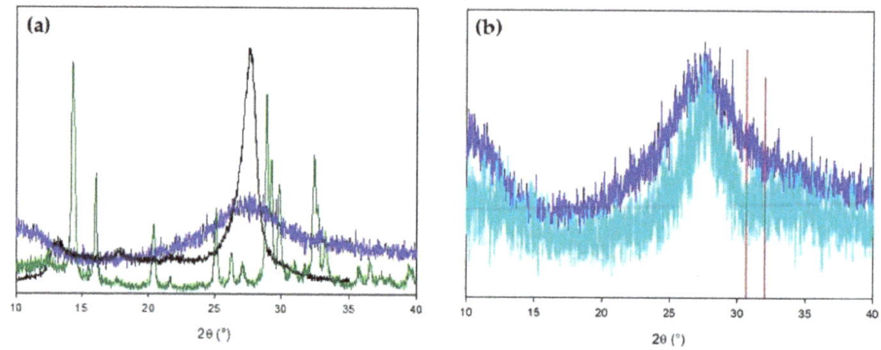

Figure 5. XRD patterns collected on (**a**) $DMASnBr_3$ (green), g-C_3N_4 (black) and as-prepared $DMASnBr_3$/g-C_3N_4 (blue); (**b**) the composite after three irradiations (azure) compared to the as-prepared one (blue) and reference metallic tin (red).

Although in the XRD pattern of the as-synthesized catalyst the fingerprints of the perovskite are not visible, a broad signal centered on the principal peak of the carbon nitride that stabilizes an amorphous structure of the perovskite is present, as previously discussed [13]. Superimposable patterns were observed on both fresh and recycled materials (see Figure 5b); moreover, no peaks related to metallic Sn were found after catalysis (instead observed working in TEOA solution, data not shown). These findings support both the integrity of the hybrid material and the maintenance of the perovskite in the photocatalyst. Thus, this can be utilized for at least three consecutive irradiations, as a proof of photo-chemical stability that strengthens the applicability of $DMASnBr_3$/g-C_3N_4 for a potential practical use.

Such new outcomes prompted us to test the composite in the presence of a polysaccharide sacrificial substance, and starch was chosen as the representative bio-polymer because of its low cost, large availability and attractiveness as bio-oxygenated for H_2 photogeneration [5,24].

Hydrogen gas was effectively produced from starch solution (see Table 4), at a concentration of 4.5 g L^{-1} (the maximum solubility [5]), although at lower amount compared to glucose. This is justified considering that that the rate of hydrogen evolution from water decreases with increasing molecular weights and structure complexity of carbohydrates [15–17,24]. Additionally, since the VB holes are the oxidizing species of the present catalytic system, an intimate contact between the sacrificial substrate and the catalyst surface is certainly required. This could be more favored with monosaccharides than polysaccharides with a branched skeleton such as starch, which realistically has slower mass transfer kinetics from solution to the composite surface sites [16,24].

Such new findings highlight that $DMASnBr_3$/g-C_3N_4 is able to carry forward H_2 photoproduction exploiting the sacrificial role of a raw polysaccharide such as soluble starch, which thus can be employed directly without any pre-treatment, e.g., enzymatic hydrolysis or microwave-assisted hydroxylation [26,27].

3. Materials and Methods

Preparation and characterization of the two new composites 5% PEA_2SnBr_4/g-C_3N_4 and 33% $DMASnBr_3$/g-C_3N_4 was reported by these authors in very recent works [12,13].

Surface area measurements were carried out by the Brunauer, Emmett and Teller (B.E.T.) single point method using a Flowsorb II 2300 (Micromeritics, Norcross, GA, USA) apparatus. The sample was accurately weighed and degassed at 80 °C for 1.5 h, under a continuous stream of a N_2-He (30:70) mixture, and then it was put in liquid nitrogen for gas adsorption. Evonik Aeroxide® P25 TiO_2 (10–50 nm, 60.8 m^2 g^{-1} [25]) was supplied by Evonik Industries AG (Hanau, Germany). H_2 evolution experiments were conducted in distilled water containing 0.025-0.2 M glucose (99.9%, Carlo Erba

Reagents) or 4.5 g L^{-1} soluble starch (ACS reagent, Carlo Erba Reagents), irradiated in Pyrex glass containers [28]. After addition of the catalyst (0.028, 0.5, 2 g L^{-1}), the suspension was deoxygenated by Ar bubbling (20 min) and irradiated for 6 h, under magnetic stirring. When using Pt as the co-catalyst, chloroplatinic acid (H_2PtCl_6, 38% Pt basis) from Sigma-Aldrich (Milan, Italy) was employed as the metal source. Since Pt is in situ photodeposited on the catalyst surface, after Ar bubbling, a small volume from an H_2PtCl_6 aqueous solution (0.15, 3 or 15 g L^{-1}) was added using a 10–100 μL micropipette to the catalyst suspension, directly in the photoreactor. This was closed with sleeve stopper septa and was irradiated, as described in the following, achieving simultaneous Pt deposition and H_2 production [23,28–30]. Irradiation was done under simulated solar light using a Solar Box 1500e (CO.FO.ME.GRA S.r.l., Milan, Italy) set at a power factor of 500 W m^{-2}, and equipped with UV outdoor filter made of IR-treated soda lime glass. The apparent photon flux, measured as previously described [5], was 1.53×10^{-7} photons moles s^{-1}. The headspace H_2 was quantified by gas chromatography coupled with thermal conductivity detection (GC-TCD) [28]. The results obtained relative to H_2 evolution are shown in the paper as micromoles of gas per gram of catalyst per hour (μmoles g^{-1} h^{-1}, HER). Reproducibility was appraised on the RSD% from four independent photoproduction runs. For catalyst recycling tests, the catalyst was recovered by filtration (0.45 μm nylon membrane), washed with plenty of distilled water, left drying at room temperature for several weeks in the dark, and finally re-used under the optimized conditions. Room temperature Cu-radiation XRD spectra were acquired by a Bruker D8 diffractometer (Billerica, MA, USA).

4. Conclusions

Lab-scale tests proved the superior photocatalytic activity of $DMASnBr_3$/g-C_3N_4 against PEA_2SnBr_4/g-C_3N_4 for H_2 evolution from glucose solution, under simulated solar light, using Pt as a co-catalyst. The systematic optimization of the experimental conditions by a DoE provided HER higher than 900 μmoles g^{-1} h^{-1} using the former composite. The results herein collected highlight the importance to work under selected conditions to maximize the H_2 yield while reducing the use of catalyst and co-catalyst. The comparison with P25 TiO_2 strengthened the catalytic efficiency of $DMASnBr_3$/g-C_3N_4, while avoiding the handling of nano-sized powders. Notably, a rewarding H_2 formation (around 140 μmoles g^{-1} h^{-1}) was gained, also avoiding any metal deposition. $DMASnBr_3$/g-C_3N_4 showed good photochemical stability to be utilized for two consecutive runs without any loss of efficiency and with a decrease of 25% in the third cycle, as verified by H_2 evolution measurement and XRD analysis. The system turned out to work also in raw starch solution as sustainable, model sacrificial biomass, providing ca. 150 μmoles g^{-1} h^{-1}. These outcomes corroborate the potential application of this new class of photocatalysts for clean energy retrieval under sustainable conditions and deserve further investigation to move the photoproduction experiments on sugar-rich wastewaters under natural solar light.

Author Contributions: Conceptualization, A.S., L.M. and A.P.; validation, A.S. and A.P.; formal analysis, A.S.; investigation, A.S. and L.R.; data curation, A.S.; writing—original draft preparation, A.S.; writing—review and editing, L.M., D.D. and A.P.; visualization, A.S. and L.R.; supervision, A.S. and A.P.; funding acquisition, A.P. All authors have read and agreed to the published version of the manuscript.

Funding: This research received no external funding.

Acknowledgments: The Authors acknowledge Giuseppina Sandri (Department of Drug Sciences, University of Pavia) for surface area measurements.

Conflicts of Interest: The authors declare no conflict of interest.

References

1. Wen, J.; Xie, J.; Chen, X.; Li, X. A review on g-C_3N_4-based photocatalysts. *Appl. Surf. Sci.* **2017**, *391*, 72–123. [CrossRef]
2. Nasir, M.S.; Yang, G.; Ayub, I.; Wang, S.; Wang, L.; Wang, X.; Yan, W.; Peng, S.; Ramakarishna, S. Recent development in graphitic carbon nitride based photocatalysis for hydrogen generation. *Appl. Catal. B* **2019**, *257*, 117855. [CrossRef]
3. Lam, S.S.; Nguyen, V.-H.; Dinh, M.T.N.; Khieu, D.Q.; La, D.D.; Nguyen, H.T.; Vo, D.V.N.; Xia, C.; Varma, R.S.; Shokouhimehr, M.; et al. Mainstrem avenues for boosting graphitic carbon nitride efficiency: Towards enhanced solar light-driven photocatalytic hydrogen production and environmental remediation. *J. Mater. Chem. A* **2020**, *8*, 10571–10603. [CrossRef]
4. Speltini, A.; Scalabrini, A.; Maraschi, F.; Sturini, M.; Pisanu, A.; Malavasi, L.; Profumo, A. Improved photocatalytic H_2 production assisted by aqueous glucose biomass by oxidized g-C_3N_4. *Int. J. Hydrogen Energy* **2018**, *43*, 14925–14933. [CrossRef]
5. Speltini, A.; Gualco, F.; Maraschi, F.; Sturini, M.; Dondi, D.; Malavasi, L.; Profumo, A. Photocatalytic hydrogen evolution assisted by aqueous (waste)biomass under simulated solar light: Oxidized g-C_3N_4 vs. P25 titanium dioxide. *Int. J. Hydrogen Energy* **2019**, *44*, 4072–4078. [CrossRef]
6. Zhang, L.; Zhang, J.; Xia, Y.; Xun, M.; Chen, H.; Liu, X.; Yin, X. Metal-free carbon quantum dots implant graphitic carbon nitride: Enhanced photocatalytic dye wastewater purification with simultaneous hydrogen production. *Int. J. Mol. Sci.* **2020**, *21*, 1052. [CrossRef]
7. Huang, H.; Pradhan, B.; Hofkens, J.; Roeffaers, M.B.; Steele, J.A. Solar-driven metal halide perovskite photocatalysis: Design, stability, and performance. *ACS Energy Lett.* **2020**, *5*, 1107–1123. [CrossRef]
8. Romani, L.; Malavasi, L. Solar-driven hydrogen generation by metal halide perovskites: Materials, approaches, and mechanistic view. *ACS Omega* **2020**, (in press). [CrossRef]
9. Stanks, S.D.; Snaith, H.L. Metal-halide perovskites for photovoltaic and light-emitting devices. *Nat. Nanotec.* **2015**, *10*, 391–402.
10. Pu, Y.-C.; Fan, H.-C.; Liu, T.-W.; Chen, J.-W. Methylamine lead bromide perovskite/protonated graphitic carbon nitride nanocomposites: Interfacial charge carrier dynamics and photocatalysis. *J. Mater. Chem. A* **2017**, *5*, 25438–25449. [CrossRef]
11. Li, Z.; Wu, S.; Zhang, J.; Yuan, Y.; Wang, Z.; Zhu, Z. Improving photovoltaic performance using perovskite/surface-modified graphitic carbon nitride heterojunction. *Sol. RRL* **2020**, *4*, 1900413. [CrossRef]
12. Romani, L.; Bala, A.; Kumar, V.; Speltini, A.; Milella, A.; Fracassi, F.; Listorti, A.; Profumo, A.; Malavasi, L. PEA_2SnBr_4: A water-stable lead-free two-dimensional perovskite and demonstration of its use as co-catalyst in hydrogen photogeneration and organic-dye degradation. *J. Mater. Chem. C* **2020**, *8*, 9189–9194. [CrossRef]
13. Romani, L.; Speltini., A.; Ambrosio, F.; Mosconi, E.; Profumo, A.; Marelli, M.; Margadonna, S.; Milella, A.; Fracassi, F.; Listorti, A.; et al. Water-stable $DMASnBr_3$ lead-free perovskite for effective solar-driven photocatalysis. *Angew. Chem. Int. Ed.* **2020**. [CrossRef]
14. Pisanu, A.; Speltini, A.; Quadrelli, P.; Drera, G.; Sangaletti, L.; Malavasi, L. Enhanced air-stability of Sn-based hybrid perovskites induced by dimethylammonium (DMA): Synthesis, characterization, aging and hydrogen photogeneration of the $MA_{1-x}DMA_xSnBr_3$ system. *J. Mater. Chem. C* **2019**, *7*, 7020–7026. [CrossRef]
15. Leung, D.Y.C.; Fu, X.; Wang, C.; Ni, M.; Leung, M.K.H.; Wang, X.; Fu, X. Hydrogen production over titania based photocatalysts. *ChemSusChem* **2010**, *3*, 681–694. [CrossRef]
16. Silva, C.G.; Sampaio, M.J.; Marques, R.R.N.; Ferreira, L.A.; Tavares, P.B.; Silva, A.M.T.; Faria, J.L. Photocatalytic production of hydrogen from methanol and saccharides using carbon nanotube-TiO_2 catalysts. *Appl. Catal. B Environ.* **2015**, *178*, 82–90. [CrossRef]
17. Fu, X.; Long, J.; Wang, X.; Leung, D.Y.C.; Ding, Z.; Wu, L.; Zhang, Z.; Li, Z.; Fu, X. Photocatalytic reforming of biomass: A systematic study of hydrogen evolution from glucose solution. *Int. J. Hydrogen Energy* **2008**, *33*, 6484–6491. [CrossRef]
18. Kisch, H. On the problem of comparing rates or apparent quantum yields in heterogeneous photocatalysis. *Angew. Chem. Int. Ed.* **2010**, *49*, 9588–9589. [CrossRef] [PubMed]
19. Speltini, A.; Sturini, M.; Maraschi, F.; Dondi, D.; Serra, A.; Profumo, A.; Buttafava, A.; Albini, A. Swine sewage as sacrificial biomass for photocatalytic hydrogen gas production: Explorative study. *Int. J. Hydrogen Energy* **2014**, *39*, 11433–11440. [CrossRef]

20. Speltini, A.; Sturini, M.; Maraschi, F.; Dondi, D.; Fisogni, G.; Annovazzi, E.; Profumo, A.; Buttafava, A. Evaluation of UV-A and solar light photocatalytic hydrogen gas evolution from olive mill wastewater. *Int. J. Hydrogen Energy* **2015**, *40*, 4303–4310. [CrossRef]
21. Baranowski, M.; Plochocka, P. Excitons in metal-halide perovskites. *Adv. Energy Mater.* **2020**, *10*, 1903659. [CrossRef]
22. Bowker, M. Photocatalytic hydrogen production and oxygenate photoreforming. *Catal. Lett.* **2012**, *142*, 923–929. [CrossRef]
23. Zhang, J.-H.; Wei, M.-J.; Wei, Z.-W.; Pan, M.; Su, C.-Y. Ultrathin graphitic carbon nitride nanosheets for photocatalytic hydrogen evolution. *ACS Appl. Nano Mater.* **2020**, *3*, 1010–1018. [CrossRef]
24. Kennedy, J.; Bahruji, H.; Bowker, M.; Davies, P.R.; Bouleghlimat, E.; Issarapanacheewin, S. Hydrogen generation by photocatalytic reforming of potential biofuels: Polyols, cyclic alcohols, and saccharides. *J. Photochem. Photobiol. Chem.* **2018**, *356*, 451–456. [CrossRef]
25. Jiang, X.; Manawan, M.; Feng, T.; Qian, R.; Zhao, T.; Zhou, G.; Kong, F.; Wang, Q.; Dai, S.; Pan, J.H. Anatase and rutile in evonik aeroxide P25: Heterojunctioned or individual nanoparticles? *Catal. Today* **2018**, *300*, 12–17. [CrossRef]
26. Li, K.L.; Xia, L.X.; Li, J.; Pang, J.; Cao, G.Y.; Xi, Z.W. Salt-assisted acid hydrolysis of starch to D-glucose under microwave irradiation. *Carbohydr. Res.* **2001**, *331*, 9–12.
27. Wang, D.; Hou, F.; Ma, X.; Chen, W.; Yan, L.; Ding, T.; Ye, X.; Liu, D. Study on the mechanism of ultrasound-accelerated enzymatic hydrolysis of starch: Analysis of ultrasound effect on different objects. *Int. J. Biol. Macromol.* **2020**, *148*, 493–500. [CrossRef] [PubMed]
28. Speltini, A.; Sturini, M.; Dondi, D.; Annovazzi, E.; Maraschi, F.; Caratto, V.; Profumo, A.; Buttafava, A. Sunlight-promoted photocatalytic hydrogen gas evolution from water-suspended cellulose: Systematic study. *Photochem. Photobiol. Sci.* **2014**, *13*, 1410–1419. [CrossRef]
29. Pisanu, A.; Speltini, A.; Vigani, B.; Ferrari, F.; Mannini, M.; Calisi, N.; Cortigiani, B.; Caneschi, A.; Quadrelli, P.; Profumo, A.; et al. Enhanced hydrogen photogeneration by bulk g-C_3N_4 through a simple and efficient oxidation route. *Dalton Trans.* **2018**, *47*, 6772–6778. [CrossRef]
30. Sun, H.; Zhou, X.; Zhang, H.; Tu, W. An efficient exfoliation method to obtain graphitic carbon nitride nanosheets with superior visible-light photocatalytic activity. *Int. J. Hydrogen Energy* **2017**, *42*, 7930–7937. [CrossRef]

Publisher's Note: MDPI stays neutral with regard to jurisdictional claims in published maps and institutional affiliations.

catalysts

Review

Two-Dimensional Materials and Composites as Potential Water Splitting Photocatalysts: A Review

Zubia Saleem, Erum Pervaiz *, M. Usman Yousaf and M. Bilal Khan Niazi

Department of Chemical Engineering, School of Chemical & Materials Engineering (SCME), National University of Sciences & Technology (NUST), Sector H-12 Islamabad, Islamabad 44000, Pakistan; zubia.saleem92@gmail.com (Z.S.); usman.yousaf93@gmail.com (M.U.Y.); m.b.k.niazi@scme.nust.edu.pk (M.B.K.N.)

* Correspondence: erum.pervaiz@scme.nust.edu.pk; Tel.: +92-51-9085-5113

Received: 28 November 2019; Accepted: 18 January 2020; Published: 24 April 2020

Abstract: Hydrogen production via water dissociation under exposure to sunlight has emanated as an environmentally friendly, highly productive and expedient process to overcome the energy production and consumption gap, while evading the challenges of fossil fuel depletion and ecological contamination. Various classes of materials are being explored as viable photocatalysts to achieve this purpose, among which, the two-dimensional materials have emerged as prominent candidates, having the intrinsic advantages of visible light sensitivity; structural and chemical tuneability; extensively exposed surface area; and flexibility to form composites and heterostructures. In an abridged manner, the common types of 2D photocatalysts, their position as potential contenders in photocatalytic processes, their derivatives and their modifications are described herein, as it all applies to achieving the coveted chemical and physical properties by fine-tuning the synthesis techniques, precursor ingredients and nano-structural alterations.

Keywords: water splitting; photocatalyst; graphene oxide; phosphorene; graphitic carbon nitride; MOFs

1. Introduction

Today, energy is one of the utmost critical environmental, economic and political issues due to the contributing factors of continuously increasing global population, escalating urbanization and growing energy consumption. These dynamics are compelling researchers and front-runners in political, social, environmental and industrial fields towards a shift from conventional energy production methods towards more innovative, sustainable, efficient, safe and environmentally benign technologies and materials. The total energy consumption of the entire globe is 18.5 TW per year currently, which is expected to rise to 40.8 TW by 2050 [1]. Approximately 85% of total energy usage comes from fossil fuel. It is estimated that there are sufficient fossil fuel resources to meet global energy demands for several centuries [2]. However, irreversible climate change due to generation of greenhouse gases, depletion of fossil fuel reserves, geopolitical conflicts and other environmental concerns are continuous threats for secure and sustainable energy supply. Therefore, the world needs to devise advanced, sustainable methods and sources competitive to fossil fuels in terms of efficiency and economics while ensuring minimal environmental damage [3]. The sun is an everlasting source of inexhaustible energy with which to produce hydrogen, which proves to be potential candidate to meet secure energy demands due to environmental friendliness, high energy value and the possibility of onsite production [4]. The amount of solar energy reaching earth is around 9.5×10^{16} J/s; that is 10^4 times the total usage by entire human population, endorsing its candidature for potential sustainable energy sources [5]. The major research concern for harnessing solar energy is the production of solar fuel. Conventional technologies of solar energy production suffer from many downsides, such as the limited storage capacity for the energy, the intermittent availability in certain areas and the economic feasibility of the process. The steam reforming process is traditionally used for hydrogen production from fossil fuels, which is not considered as sustainable in the long haul due to massive carbon dioxide emission and depleting resource trends. Catalytic steam reforming using fossil fuels is virtually the only industrial process responsible for producing the entire hydrogen supply of the world today, with electrolysis contributing only a meager 4% via water electrolysis. There is a growing interest in generating hydrogen from sustainable and renewable resources using environmentally friendly methods to deliver inexpensive and clean energy [6]. Photocatalytic water splitting activity is one of the key research areas for onsite hydrogen production and storage [7–16]. It is one of the emerging process for harnessing solar energy into fuel in the form of hydrogen without any use of fossil fuels or emission of carbon dioxide. There is a growing interest in the field of semiconductor photocatalysis [17] which could be used for water dissociation into constituent oxygen and hydrogen gases using solar energy, which is abundantly available all over the world. Solar energy conversion in the form of hydrogen via photocatalytic water splitting has become an attractive subject [18].

The prerequisites to an energy efficient, cost effective and operationally feasible photocatalyst are the ability to effectively harvest and utilize photons in the visible light region to generate electron-hole pairs; lower band gap energies; unimpeded mobility of charge carriers; a low recombination rate of electron-hole pairs; and a greater surface area, resulting in a higher number of exposed active sites. The earliest discoveries fulfilling these requirements comprised granular semi-conducting materials and their derivatives, such as ZnO and TiO_2 [19]. But these conventional photocatalysts have several shortcomings, such as:

(1) Their higher band gap energy, making the materials suitable for mainly UV light absorption instead of visible light [20];
(2) Only having the ability to catalyze either water oxidation or reduction at a time, leading to their unsuitability to act as dual function overall water splitting catalysts [21];
(3) Their higher charge recombination rates in bulk medium and on the surface, resulting in lower activity [22];
(4) Inaccessibility of active sites lying in bulk material [23].

Further research and development in this regard led to the conclusion that the nano-materials are better photocatalysts because of their high surface area property [24,25]. Among nanomaterials, 2D materials are promising photocatalyst alternatives due to their sheet-like structures; electrical, thermal and mechanical properties; the operational advantages of being easily handleable; having the ability to form integrated composites with highly photoactive co-catalysts; and having higher surface to volume ratios. The bi-layer and mono-layer 2D nanocrystals and nanosheets provide the basis for highly exposed active sites and faster charge transfer schemes [26], while the doped precious and photo-active metals and their related compounds offer the added compensation of higher light absorption capacity and lower band gap energy [27,28].

1.1. Photocatalysis: Theoretical Digest

Hydrogen production by photocatalytic water splitting has proven to be a superlative solution for resource and economics related problems in novel hydrogen production methods. The process is centered on a semiconducting catalyst material and sunlight to convert the light energy to chemical energy, also termed "artificial photosynthesis." The photons of light having higher energy than the energy gap of semiconducting material, when irradiated on the photocatalyst material, results in photo generated charge carriers usually termed the electron hole pair. This generation of electron-hole pairs in the conduction band and valence band acts as the foundation for redox reactions, as indicated below [29,30]:

Overall:

$$2H_2O\ (l) \xrightarrow{Photon\ energy>1.23\ eV} O_2(g) + 2H_2\ (g)$$

$$E^o = 1.23\ V.$$

The oxidation and reduction half reactions at anode and cathode, respectively, under acidic (low pH) solutions proceed as:

Oxidation:

$$2H_2O\ (l) + 4h^+ \rightarrow O_2(g) + 4H^+(aq)$$

$$E^o_{oxidation} = -1.23\ V.$$

Reduction:

$$2H^+ + 2e^- \rightarrow O_2(g) + 2H_2\ (g)$$

$$E^o_{reduction} = 0\ V.$$

Under basic (high pH) conditions, the chemical reactions are represented as:

Oxidation:

$$2OH^-(aq) \rightarrow \frac{1}{2}\, O_2(g) + H_2O\ (l) + 2e^-\ .$$

Reduction:

$$2H_2O\ (l) + 2e^- \rightarrow H_2\ (g) + 2OH^-.$$

The holes generated in the valence band resultantly are responsible for oxidation of adsorbed species (water in this case) releasing oxygen, and the agitated electrons from the conduction band are the reducing agents to produce hydrogen from in-contact water molecules as shown in Figure 1.

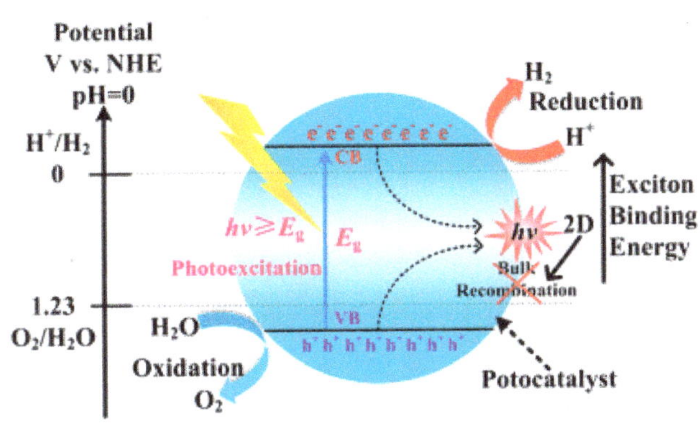

Figure 1. Schematic illustration of basic processes in photocatalytic water splitting. (Reprinted with permission from ACS Catal. 2018, 8, 3, 2253–2276, Publication Date: 30 January 2018. Copyright (2018) American Chemical Society [23]).

The production of charge carriers, such as electron and hole ones, is a fundamental step for all photocatalytic processes; other crucial factors include the width of material band gap and the energy levels of both conduction and valence bands of the material. For a feasible reaction to occur, the conduction band of the electron acceptor species should be more positive than the conductive band of the semiconductor photocatalyst, whereas the valence band of the electron donor type must be more negative than the semiconductor's valence band, and the minimum band gap must be greater than 1.23 eV [31]. As a result of the interface electron transfer process, the redox reaction takes place and the photolysis of water molecules occur.

1.2. Limitations of Photocatalytic Water Splitting

1. To catalyze the splitting of water at the interface of electrolyte and electrode, the charge carriers are required to be transferred to catalyst surface once the electron-hole pairs are created. The quick

recombination of photo-generated electron-hole pairs (Figure 2) releasing heat or photon energy before they can catalyze the redox reactions is a major challenge in this step, requiring a high degree of crystallinity [32].

2. The predilection for semiconductor materials to work under the ultraviolet (UV) light is another major challenge, as only about 4% of solar energy is comprised of UV light. It is advantageous for photocatalysts to work under visible light, which requires the band gap to be in visible range.
3. Photo-corrosion and catalyst decay are also among the limitations of the photocatalytic splitting of water. TiO_2, ZrO_2, $KTaO_3$, $SrTiO_3$ and $BiVO_4$ are among the notable contenders for photocatalytic water splitting because of having band gaps around 1.23 eV. Typical sulfide-based photocatalysts, such as cadmium sulfide (CdS), have a tendency to undergo decay under operating conditions due to oxidation of sulfide into elemental sulfur at the same potentials that are used for water splitting, requiring the use of certain sacrificial reagents, e.g., sodium sulfide, to control any sulfur lost [33].

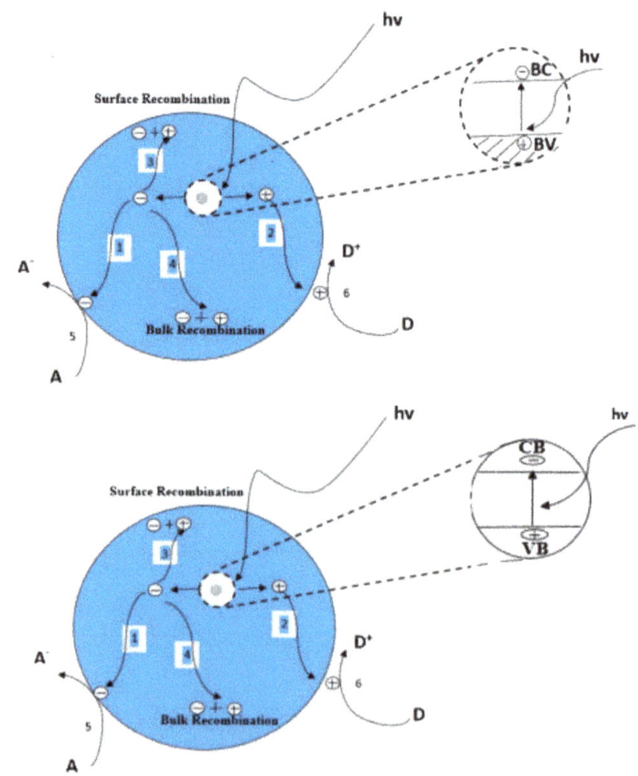

Figure 2. Schematic of charge carrier dynamics in a semiconductor: one electron transported to the surface; two holes transported to the surface; three surface recombinations; four bulk recombinations; five electron transfers to respective acceptor molecules; and six hole trappings by a donor molecule. (Reprinted with permission from [34]. Copyright (2019), Elsevier).

2. Two Dimensional Materials as Photocatalysts for Water Splitting

A good photocatalyst must possess the characteristics of photo-activity, thermal stability, photo-stability, chemical inertness, low cost, non-toxicity and post-reaction recoverability [35]. To tackle the environmental pollution factor, metal-free catalysts are always the first priority for synthesize with maximum efficiency. The first and second generation of catalysts include combinations of inorganic materials [36] and various oxides and sulfides of metals, but the main problem of these materials is the toxic and corrosive nature [33,37]. Among various other photocatalyst materials available, 2D materials are promising candidates for the water splitting applications, because of merits, such as:

(1) The adjustable number of layers, so that the band gap and photo-absorption can be tuned [38].
(2) The sheet-like ultra-thin structure facilitating the charge transport towards surface, resulting in lower recombination rate [39].
(3) The enhanced specific surface area leading to greater exposure of active sites on surface and better catalytic performance [40].

The fundamental strategies to synthesizing 2D nanosheets comprise top-down and bottom-up methods, which vary on the basis of starting precursors and treatment methods also shown in Figure 3.

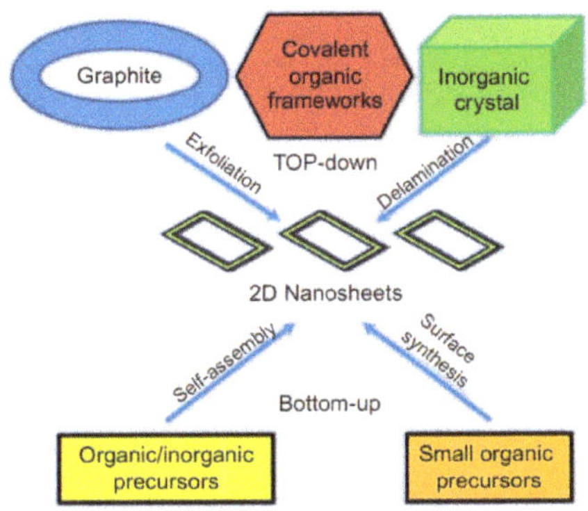

Figure 3. Schematic illustration of top-down and bottom-up strategies toward the synthesis of 2D nanosheets. (Reprinted with permission from [41]. Copyright (2015), Elsevier).

Till now, various 2D materials have been discovered by researchers, such as graphene, graphitic carbon nitride, black phosphorus/phosphorene and hexagonal boron nitride. All these materials have proven potential for water splitting activity by photocatalysis [42]. 2D materials show unusual structural, electronic and optical properties [43]. However, some limitations have been observed in terms of reaction completion due to the recombination rate of the electrons and holes which carry out reactions. In the search for alternatives to metal-based systems, metal-free elementals and compound photocatalysts have recently been developed. Red-P, alpha-sulfur and boron are notable examples reported in the literature [44]. However, these elemental photocatalysts exhibit low photocatalytic activity. There are various techniques that researchers employ to enhance activity majorly, including doping, sensitization, tuning morphology and making heterostructures with other materials [45]. The search for a non-toxic, inexpensive, corrosion resistant photocatalyst with a focus on the design and efficiency of photocatalyst, while assessing the activity for economical and scalable hydrogen generation, is ongoing. Table 1 summarizes the discussed literature in a comprehensive manner. The main focus of this paper is to go over 2D materials, their modifications and the different techniques that have been employed to enhance their photocatalytic water splitting performances. Said materials include, but are not limited to, graphitic carbon nitride, graphene, black phosphorus, metal phosphides and metal organic frameworks. Figure 4 presents the bandgap energies of all potential 2-D materials.

Table 1. A summary of recent studies on ultra-thin 2D photocatalysts. (Adapted with permission from [46], copyright (2017); Wiley and [47], copyright (2019); Nanotechnology, NCBI and [48], copyright (2019), Springer) (Unless otherwise specified).

Photocatalyst	Thickness (nm)	Bandgap (eV)	Sacrificial Agent	Co-Catalyst	Light Source	Type of Reaction	Activity (µmol/g·h)	Ref.
Pore rich WO_3 ultra-thin nanosheets	3.8	2.89, CB −0.40, VB 2.48	0.5 mol/L Na_2SO_4	-	300 W Xe lamp	OER	Photocurrent density 2.14 mA/cm^2 at 1.0 V	[49]
O-vacancy-rich In_2O_3 nanosheets	0.9	2.18	-	-	300 W Xe lamp with 420 nm cutoff filter	OER	Photocurrent density 1.73 mA/cm^2 at 0.576 V	[50]
Surface atomic SnS sheets	0.57	1.47	0.5 mol/L Na_2SO_4	-	300 W Xe lamp with 420 nm cutoff filter	Overall	Photocurrent density of 5.27 mA/cm^2 at 0.8 V	[51]
Single layer SnS_2	0.61	2.23	-	-	300 W Xe lamp with 420 nm cutoff filter	Overall	Photocurrent density of 2.75 mA/cm^2 at 1.0 V	[52]
ZnSe	0.85	3.5	0.5 mol/L Na_2SO_4	-	300 W Xe lamp	Overall	Photocurrent density of 2.14 mA/cm^2 at 0.72 V	[53]
Cu_2O	0.62	1.92	0.5 mol/L Na_2SO_4	-	300 W Xe lamp with 420 nm cutoff filter	HER	Photocurrent density of 3.98 mA/cm^2 at −1.0 V	
HNb_3O_8	1.3	3.68	10 vol% TEOA	1 wt% Pt	125 W Hg lamp	HER	≈610	[54]
$HNbWO_6$	1.8–2.0	3.13, CB −0.68, VB 2.45	TEOA	1 wt% Pt	300 W Xe lamp	HER	1986.25	[55]
$SnNb_2O_6$	≈3	2.43	20 vol% lactic acid	0.3 wt% Pt	300 W Xe lamp with 400 nm cutoff filter	HER	264	[9]
CoOOH	1.5	2.4	0.5 M Na_2SO_3	-	300 W Xe lamp	HER	1200	[56]
CdS	≈4	≈2.86	0.25 M Na_2S, 0.25 M Na_2SO_3	-	300 W Xe lamp with 420 nm cutoff filter	HER	41,100	[57]
O-doped $ZnIn_2S_4$	6	2.07, CB −1.34, VB 0.73	0.25 M Na_2SO_3, 0.35 M Na_2S	-	300 W Xe lamp with 420 nm cutoff filter	HER	2120	[58]
g-C_3N_4	≈2	2.97	10 vol% TEOA	6 wt% Pt	300 W Xe lamp	HER	170.5	[39]
g-C_3N_4	≈2	2.65	10 vol% TEOA	3 wt% Pt	300 W Xe lamp with 420 nm cutoff filter	HER	1860	[59]
$MoS_2/Bi_{12}O_{17}Cl_2$	0.71 of $Bi_{12}O_{17}Cl_2$	≈2.5	0.3 M Ascorbic acid	-	300 W Xe lamp with 420 nm cutoff filter	HER	33,000	[60]
Rh-doped calcium niobate	≈2.8–3.0	≈1.9	10 vol% CH_3OH	Rh	500 W Xe lamp	HER	76,960	[61]
C-BN	3–4	2.72, CB −1.24, VB 1.48	10 vol% TEOA	1 wt% Pt	300 W Xe lamp	HER	≈920	[62]
Rh single atoms/TiO_2	0.7	-	20 vol% CH_3OH	Single atom Rh	500 W Xe lamp	HER	2550	[63]
$MoSe_2/ZnIn_2S_4$	2.5 of $ZnIn_2S_4$	≈2.75	10 vol% Lactic acid	1 wt% $MoSe_2$	300 W Xe lamp with 400 nm cutoff filter	HER	6454	[64]

Table 1. *Cont.*

Photocatalyst	Thickness (nm)	Bandgap (eV)	Sacrificial Agent	Co-Catalyst	Light Source	Type of Reaction	Activity (μmol/g·h)	Ref.
$Ti_3C_2/g\text{-}C_3N_4$	1.7–2.6 of Ti_3C_2	-	10 vol% TEOA	3 wt% Pt, 3 wt% Ti_3C_2	200 W Hg lamp $\lambda > 400$ nm	HER	72.3, 0.81% quantum efficiency at 400 nm	[65]
10% $Fe_2O_3/g\text{-}C_3N_4$ (1.0 wt% Pt)	21 Fe_2O_3	2.8 of $g\text{-}C_3N_4$, 2.1 of Fe2O3	15 vol% TEOA	1 wt% Pt	350 W xenon lamp $\lambda > 420$ nm	HER	398.0	[66]
$WO_3/ZnIn_2S_4$	-	2.4 of $ZnIn_2S_4$	0.25 M Na_2SO_3/0.35 M Na_2S	-	300 W xenon lamp $\lambda > 420$ nm	HER	2202.9	[67]
$Cu_2S/Zn_{0.67}Cd_{0.33}S$	5	2.28–2.5	0.1 M Na_2S/Na_2SO_3	Cu_2S	300 W xenon lamp $\lambda > 420$ nm	HER	15270, quantum efficiency 18.15% at 420 nm	[68]
15% $WO_3/g\text{-}C_3N_4$	2.5–3.5	2.77 of WO_3 and 2.68 of $g\text{-}C_3N_4$	20 vol% Lactic acid	2 wt% Pt	350 W xenon lamp	HER	982	[69]
25.0% UNiMOF/$g\text{-}C_3N_4$	3.04 UNiMOF	-	10 vol% TEOA	-	300 W xenon lamp $\lambda > 420$ nm	HER	400.6, quantum efficiency 0.979% at 420 nm	[70]
0.75% $MoS_2/g\text{-}C_3N_4$	8–10 of $g\text{-}C_3N_4$	-	0.1 M TEOA	0.75 wt% MoS_2	300 W xenon lamp $\lambda > 420$ nm	HER	1155, quantum efficiency 6.8% at 420 nm nm	[71]
$MoS_2/SnNb_2O_6$ (1.0 wt% Pt)	-	2.59 of SNO, 1.8 of MoS_2	20 vol% CH_3OH	1 wt% Pt	300 W xenon lamp $\lambda > 420$ nm	HER	258	[72]
$Bi_6Fe_2Ti_3O_{18}$/BiOBr ferroelectric heterostructure	20–50	2.5 of BFTO	0.2 g $AgNO_3$	-	300 W Xe arc lamp $\lambda > 420$ nm	OER	13.8	[73]
Ultrathin Bi_3O_4Cl/BiOCl	0.799–0.961	2.6	$AgNO_3$ and $FeCl_3$	-	300 W Xe lamp with the 400 nm cut-off filter	OER	58.6	[74]
$MoSe_2/Ag_3PO_4$ Heterojunction	-	-	0.5 mol/L Na2SO4	-	white LED	OER	182	[65]
BiO_{2-x} Ultra-thin nanosheet	5–10	1.3	Methyl viologen (2:20 ratio)	-	300 W Xenon arc lamp	OER	2715.443	[75]
CoO_x/hexagonal $\alpha\text{-}Fe_2O_3$	-	2.09 of $\alpha\text{-}Fe_2O_3$	0.1 g $AgNO_3$	5 wt% CoO_x	300 W Xenon lamp with λ 400 nm cut-off filter	OER	195.19	[76]
$Co\text{-}Zn_{0.5}Cd_{0.5}S$	-	2.45	$Na_2S\text{–}Na_2SO_3$	0.5 wt% Co	visible light (>420 nm)	HER	17.36	[77]
700-CoOx-C	-	-	$[Ru(bpy)_3]^{2+}\text{–}Na_2S_2O_8$	-	visible light (>420 nm)	OER	0.039	[78]

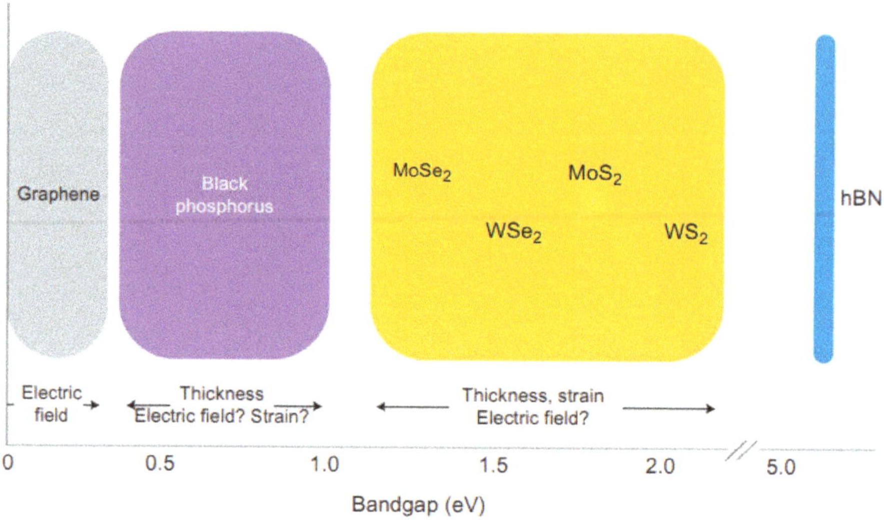

Figure 4. Bandgap energies of several layered materials used for nanoelectronics. (Reprinted with permission from [41]. Copyright (2015), Elsevier).

2.1. Graphitic Carbon Nitride and Derivatives

Graphitic carbon nitride is metal-free polymeric catalyst comprised of repeating triazine units. It is proving to be a highly efficient catalyst for photocatalytic activity with band gap of 2.56–2.70 eV. Due to highly efficient performance with an environmentally friendly nature, it is one of the most highly esteemed materials for water splitting. Due to some of its limitations, research is underway to make it more feasible and economical with improved activity. Different techniques, such as morphological alteration by changing temperature, modifying g-C_3N_4 with some metal and doping, for making heterostructures, have been investigated [79–81]. These modifications are targeted toward achieving a highly efficient catalyst by lowering the recombination rate of electrons and holes, improving value and improving environmental friendliness as much as possible [82]. 2D nanomaterials are being fabricated by facile and reliable methods: top-down ones include thermal exfoliation, chemical exfoliation, mechanical cleavage and selective etching; the bottom up approaches include chemical vapor deposition and wet synthesis methods. The most employed technique for the fabrication of 2D g-C_3N_4 is thermal treatment [83]. This facile method is easy to scale up for industrial level synthesis. However, in order to improve the photocatalytic performance, different techniques are employed and are under investigation to get maximum efficiency of a given material. Different temperature controlled synthesis techniques have been investigated to alter the morphology of the material [84]. Quan Gu investigated the morphological evolution of graphitic carbon nitride nanostructures and also evaluated their photocatalytic activities under visible light. The materials were synthesized at different temperatures, and showed different morphologies, optical properties, energy band structures and photocatalytic activity for water splitting. By changing the temperature, different structures, such as nanosheets, nanotubes and nanoparticles, were observed. BET surface area increased dramatically from 4.5 m^2/g of bulk g-C_3N_4 to 210.1 m^2/g of modified sample obtained at 540 °C. The exfoliated structures showed better activity for water splitting; i.e., 297 and 120 μmol/g·h for HER and OER respectively [85]. Zhi -An Lan modified g-C_3N_4 with bromine to investigate the change in properties. It was observed to get enhanced optical, conductive and photocatalytic properties of g-C_3N_4. The optimal catalyst CNU-Br 0.1 demonstrated more than two times higher H_2 evolution activity than pure CNU [86]. Junqing Yan fabricated g-C_3N_4/TiO_2 heterojunction by the hydrothermal method to get good production and separation of charge carriers. The fabricated TiO_2/g-C_3N_4 samples are active in the hydrogen evolution reaction of water splitting with the best performance of 700 μmol/g·h and better photo-stability [87,88]. Recently, Yuhao Yang proposed a 2D/2D heterojunction strategy in order to compensate for low efficiency of charge separation in ultra-thin g-C_3N_4 2D sheets. The high surface area interface of g-C_3N_4 nanosheets and 2D TiO_2, coupled with favorable band structure, reliability of operation and stability of TiO_2 nanosheets resulted in high H_2 evolution rates and a robust, recyclable photocatalyst [89]. Guigang Zhang conducted some experimental studies of various redox cocatalysts, reporting g-C_3N_4 modified with Pt, PtO_x, and CoO_x. The results proved that cocatalysts, especially Pt/g-C_3N_4, show potential for water splitting activity without the use of a sacrificial agent [88]. Yaping Zeng fabricated $NiTiO_3$/g-C_3N_4 heterostructured material by the sol gel method. It was observed that $NiTiO_3$ had a suitable band gap and unique photo response in visible light range. It showed three times the activity of $NiTiO_3$/g-C_3N_4 hybrid catalysts, as compared to pure g-C_3N_4 and 3 wt%, the catalysts presented optimal activity. This is ascribed to the strong light absorption in the visible region and the prolonged recombination rate of electrons and holes [90].

Xiaojie She aimed at the template free synthesis of porous ultrathin nonmetal doped g-C_3N_4 nanosheets for increased H_2 production. The formation of an ultrathin structure and the introduction of oxygen proved favorable for the enhancement of the photocatalytic performance. The average H_2 evolution rate of came out at ≈ 3786 μmol/g·h; that is five times that of bulk g-C_3N_4 photocatalyst material. This increased activity was endorsed due to more adsorption and active sites, the enhanced redox ability and improved electron transportability due to the introduction of the electrophilic groups (C-O, C=O and COOH) [91]. Mohammad W. Kadia et al., in their work, claimed to increase the efficiency of water splitting using visible light, and also increased the charge separation using mesoporous g-C_3N_4 which was coated over by WO_3. They further claim that by coating a semiconductor surface with a metal-oxide, it can capture the surface electrons, and hence can increase the charge separation. The parameters affected the WO_3/g-C_3N_4 in charge separation ability, mesoporous structure gap between valence and conduction band high surface area [92]. Recently, a step-heterojunction scheme was demonstrated by Junwei, Fu. et al. [69], whereby 2D/2D WO_3/g-C_3N_4 photocatalyst was developed by ultrasonic exfoliation of WO_3 and dual-step thermal etching of g-C_3N_4 in bulk forms. The resulting material showed 1.7 times higher H_2 production activity by barricading the charge recombination. This advancement is owed to the fact that the electronic transfer on interface of both materials from g-C_3N_4 to WO_3 balanced the higher fermi level of g-C_3N_4 and the lower fermi level of WO_3. This mechanism creates a continuous electron transfer trend and simultaneously prevents the useful charge carriers (electrons from CB of g-C_3N_4 and holes from VB of WO_3) from recombining. $KNbO_3$ has been researched extensively in terms of photocatalytic activity for water splitting, but it has low quantum efficiency, showing water splitting activity only under UV light irradiation. It has been used to enhance g-C_3N_4 activity by making heterojunction. Dongbo, Xu. et al., in their research, presented the synthesis of g-C_3N_4/$KNbO_3$ heterojunction composites by the hydrothermal method and used them for water splitting and hydrogen production using sunlight. The g-C_3N_4 and $KNbO_3$ had better performances for water splitting by more than two and 1.8 times that of their pure forms respectively [93].

Seza, A. et al. presented microwave assisted synthesis of g-C_3N_4/SnO_2 by simple pyrolysis of urea. It is the first ever synthesis of g-C_3N_4 using microwave assistance, wherein urea is converted to g-C_3N_4 nanorod, and then the uniform distribution of SnO_2 over the surface makes it a promising material for visible light photoactivity to be used for water splitting [94]. Jingran Xiao, et al., prepared g-C_3N_4 on ZnO wires, and then incorporated nano clusters of Pt over the surface. Both the g-C_3N_4 and ZnO act as photo anodes while Pt carries the charges generated; i.e., electrons generated. This showed great improvement in the photoactivity in sun light; the efficiency increased by 9.5 times that of ZnO pure and 4.5 times that of g-C_3N_4/ZnO. The recombination of electron holes further reduced by the newly formed anode with Pt over surface. Composites of ZnO and g-C_3N_4 were prepared and nano-clusteres were decorated over the surface, which eventually increased the performance [95]. Feng Guoa et al., formed another heterojunction of g-C_3N_4 with CoO; studies showed that this metal oxide decoration over the surface of graphite carbon nitrite increased the photo activities of the newly formed photocatalyst.

The formation of this hetero-conjunction catalyst was done through the "solvothermal" process. Among all of the combinations, 30 wt% CoO and g-C_3N_4 had H_2 evolution rates of 50.2 μmol/g·h and 27.8 μmol/g·h for O_2 evolution [96]. Kelin He et al. fabricated a nano-heterojunction co-catalyst system based on the doping of hexagonal Ni_3C nanoparticles produced by thermolysis at low temperature on the base g-C_3N_4 nanosheets via simple grinding method. The catalyst comprising of over 15 wt% Ni_3C content showed the hydrogen production rate of 1518 μmol/g·h when irradiated with visible light and an apparent quantum yield 116 fold greater than pure g-C_3N_4; i.e., 0.4% at 420 nm successfully showing better results than noble metal 0.5 wt% Pt/g-C_3N_4 sample. This relatively low cost, good performance, non-noble metal addition of co-catalyst enables a decrease in hydrogen production over-potential, improvement in oxidation kinetics and effective charge carrier separation [97]. Another Non-noble metal alternative with superior hydrogen evolution rate of 169 μmol/g·h has been reported by Lu Chen et al. The technique centered on the combination of Ni_3N nanoparticles and g-C_3N_4 nanosheets via a hydrothermal method and subsequent annealing step which ensures stabilization of Ni_3N structures on g-C_3N_4 and maintains the hydrogen productivity. The notable advantages of this material include the lower cost, higher throughput than 3 wt% Pt/g-C_3N_4 material, multiple cycles operation, better carrier separation and visible light operation. These properties are attributed to the optimal loading and distribution of Ni_3N active co-catalyst medium and better stability of Ni_3N phase with a minimal conversion to less active Ni or $Ni(OH)_2$ states [98]. Based on the single layer fabrication and promising scalable strategy of single layer SnS_2 nanosheets of 3-atom thickness exhibiting above 38% water splitting efficacy under visible light by Yongfu Sun et al. Shao-hua Chen et al. further investigated the prospects of coupling SnS_2 nanosheets with g-C_3N_4 to overcome the performance limitations of afore-mentioned research. The results were conformed with the intended improvements of better charge carrier separation and band alignment owing to the weak Vander Walls heterojunction between SnS_2 and g-C_3N_4 [99]. Another approach using environmentally benign materials and methods was introduced by Rongchen Shen et al., in which the hybridization of Co_2P nanoparticles formed from phosphorization technique of cobalt metal organic framework and g-C_3N_4 nanosheets. The resulting g-C_3N_4/Co_2P/K_2HPO_4 material formed by simple grinding method exhibited a 556 times superior H_2 production rate compared to simple g-C_3N_4, i.e., 556 μmol/g·h, owing to the lower hydrogen production over potentials, better light absorbance, diminished charge carrier combination tendency and provision of an alternate reaction pathway by proton-reduction [100]. A recent approach to produce cobalt phosphide (CoP)/g-C_3N_4 cocatalysts has been presented by Xiao-Jun Sun et al., involving dual-step calcination of ZIF-67. The calcination was carried out to formulate CoP nanoparticles, which were found to boost the hydrogen release rate to as high as 201.5 μmol/g·h with a meager NPs loading of 1.42 wt%. ZIF-67 was formulated by the traditional method and g-C_3N_4 by melamine thermal condensation; they were subsequently combined, ground and calcined at a high temperature. The H_2 release rate exhibited a directly proportional relation with increasing ZIF-67 concentration until 5 wt% (1.42% CoP), indicating deterrence in ability to absorb light. Moreover, the lower water contact angle of the hybrid material than those of the individual components conforms to the assumption of better water absorption ability and reduction of adsorbed proton specie. These factors, along with the notably lower charge transport resistance indicated by photocurrent experiments, contribute to the exemplary alliance and compatibility of CoP and g-C_3N_4 as photocatalysts [101]. Another methodology to incorporate graphite carbon nitride as a root material was verified by garnishing carbon based quantum dots on g-C_3N_4 nano-tubes by thermal co-polymerization method. Yang Wang et al. reported the resultant hydrogen release rate as high as 3538.3 μmol/g·h due to reduced charge recombination and greater light capturing capabilities, demonstrating 10.94% quantum yield. The co-condensation of these components encourages tube arrangement and micro-sized composite structure to stimulate transfer of photoelectrons. Introduction of C-QDs notably shifted the pure g-C_3N_4 nano-sheet structure towards nano-tube structure, because of the tough amide bonds created among C-QDs and urea precursor. Furthermore, the band gap of untreated base material was shifted from 2.70 to 2.60 eV due to the structural change and multi-reflective containment of incident light radiations in the nano-tubes [102].

Nan Wang et al., in their research, developed MnO_2 with a co-modified nanocarbon tube with a g-C_3N_4 ternary composite. Different properties were studied, and the compound was characterized by different conventional techniques. It was observed that this ternary compound shows more active photocatalysis in contrast with carbon nanotubes (CNTs)/C_3N_4. Here, CNTs capture the electrons generated when light falls over the surface of this catalyst, which resultantly increases the rate of water splitting; hence, more hydrogen is formed. Hydrogen production is increased with rate of 4067 μmol/g·h [103]. Figure 5 shows the composite g-C_3N_4/GO with MoS2 as potential photogenerated carrier separator.

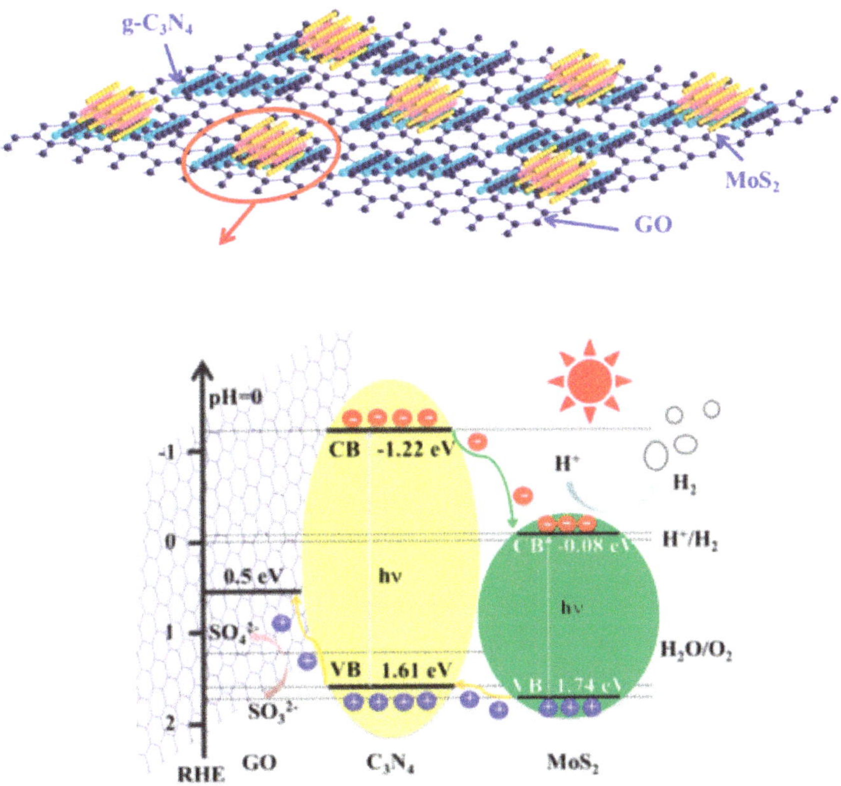

Figure 5. Schematics of MoS2/g-C_3N_4/GO composite and charge carrier separation. (Reprinted with permission from [104]. Copyright (2017), American Chemical Society).

2.2. *Graphene-Based Photocatalysts for Water Splitting*

Another metal-free photocatalyst for photocatalytic water splitting activity is graphene. It has gotten immense attention recently. In 2004, demonstration of the "sticky-tape peeling" of atomically thin sheets of carbon atoms from a mass of graphite brought a revolution in 2D materials research. This thinnest-ever material is known as "graphene" [105]. It is characterized by excellent chemical, mechanical and electrical properties [106]. For example, graphene is 100 times harder than steel; superior to copper in terms of electrical and thermal conductivity; and is flexible and transparent. Proposed future applications include uses ranging from computer chips and flexible displays to batteries and fuel cells [107].

Graphene is proving to be a potential candidate for the production of hydrogen, due its low cost and environmentally friendly nature, and shows unique charge carrier mobility properties, large specific surface area (single layer: 2650 m^2/g) [108], high transparency, structural flexibility, chemical stability and unique optoelectronic properties [109]. Graphene has been established as a viable and efficient material choice for improved catalytic operation with various other semi-conducting materials [29] due to its extended two-dimensional sp^2 hybrid carbon network framework structure. It offers excellent electron transporting and accepting characteristics, due to which it behaves as a multi-dimensional passageway for electrons, and hence, helps to effectively separate the photo-generated charge carriers. However, the possibility and efficacy of graphene to transport electron holes is yet to be explored and studied. Moreover, a great number of established studies have been concentrated on combinations of a single type of nanoparticle and graphene film, for use as photocatalysts to improve hydrogen gas production (Figure 5); but the incorporation of two or more components as co-catalysts on graphene for greater selectivity and improved performance has not been achieved. Graphene can be produced via top down or bottom up approach (Figure 6) and synthesis methodology effects the intrinsic property of the graphene. The atomically thin and flexible graphene layers can not only provide a support for dispersing metallic or oxide nanoparticles and provide a highly conductive matrix, but also can induce easier electron transfer from the conduction band of semiconductors to graphene due to large energy level offset formed at the interface, leading to an efficient charge separation.

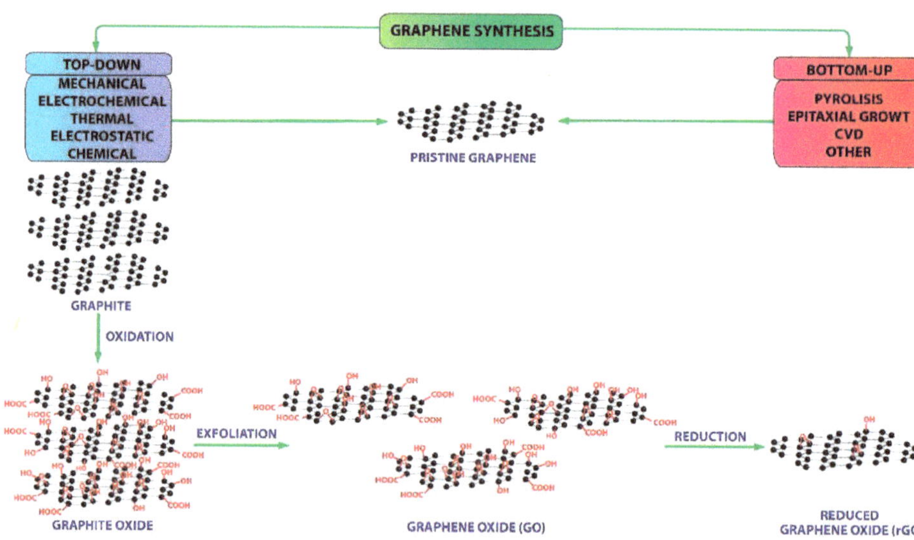

Figure 6. Top-down and bottom-up synthesis schematics of graphene. (Reprinted from [110], Copyright (2017), MDPI).

The different structures of graphene have been explored to enhance the photocatalytic performance of graphene. Graphene can be only used with a combination of other materials due to its lack of a natural band gap. Yibo Yan et al. prepared functionalized graphene quantum dots to improve performance [111]. Since the development of photocatalytic technology, TiO_2 has been a photocatalyst due to its high efficiency, low cost and good stability. TiO_2/graphene composites are currently being considered as one of the promising candidates for photocatalytic applications. The study by Angel Pérez del Pino et al. is focused on developing an inexpensive, metal-free, eco-friendly and benign photo-active catalyst while eliminating unsafe metals and reagents with superior efficiency and practicability. Laser-exposure leads to the development of defects in lattice structure, crinkles and holes, nitrogen integration, reduction of GO recognized by a drop-in oxygen atoms attached to GO and a rise in carbon content. It was further ascertained that the proliferation in N-content in GO structure results in better charge segregation, hydrogen liberation and photo-catalytic efficacy. Thus, it was logically concluded that the favorable factors for improving performance are nitrogen content assimilated in graphene and a degree of reduction, because redox reactions are favorably promoted by carbon adjacent to pyridinic-N. A hydrogen gas production rate of up to 89.29 μmol/L was recorded [112]. Shixiong Min et al. prepared TiO_2–rGO–CoO_x advanced photo-catalyst via two step hydrolysis-thermal process to separately integrate TiO_2 and CoO_x species on rGO sheets. Despite the direct non-participation of rGO as a catalyst in redox processes in water splitting, its incorporation results in noteworthy increases in gas production rates and catalytic activity due to its quality of greater electron and hole kinetics, translating to lower charge combination and transportation of electrons from TiO_2, and hole transport to CoO_x [113]. Cheng, P., and Yang, Z. et al., improved the performance of typical photocatalyst TiO_2 by combining it with graphene, which has unique properties, such as flexible structure, huge specific surface area, high transparency and electron mobility, making it a worthy contender to be combined with TiO_2 to improve its photoactivity. The experimental results specified the solvothermal reduction of graphene oxide to graphene sheets, thereby augmenting the light absorption ability of P25–GR nanocomposites and also their charge separation efficiency. Graphene played the role of photogenerated electron acceptor due to its 2D π-conjugation structure and helped as an effective transporter in the separation of electron–hole pairs. The hybrid displayed superior activity towards the evolution of hydrogen from methanol solution under the illumination caused by Xe-lamp when compared to bare P25 due to improved ability for light absorption and a lower electron–hole pair recombination rate (Figure 7). The examination of photocatalytic activity of P25–GR nanocomposites based on graphene content affirmed a mass ration of 0.5 wt% GR to P25 as the optimum, with further addition of graphene leading to a decreased photocatalytic activity [39]. Li et al. worked to enhance photo-activity of traditional TiO_2-rGO promoter and introduce an innovative method to fabricate these materials which is environmentally friendly, single-step, intense chemical-free and upgrades the gas production capability of the system by dropping band gap energy. The absence of surface Ti^{3+} species and its presence in the bulk of the sample results in better performance with respect to H_2 liberation, creation of an alternate e-conduit, modifying the reaction from UV to the visible spectrum and absorbing charge transporters. Materialization of titanium–carbon bonds leads to reducing the band gap of titanium oxide, expanding the photosensitivity range, declining the charge conduction resistance and facilitating conduction between rGO and TiO_2 (Figure 8). Hydrogen productivity grew with increasing rGO content and laser exposure duration. Peak H_2 formation rate recorded was 16000 μmol/g·h and solar transformation efficiency was 14.3%, which were the subsequent outcomes of strongly coupled TiO_2 and rGO components [91].

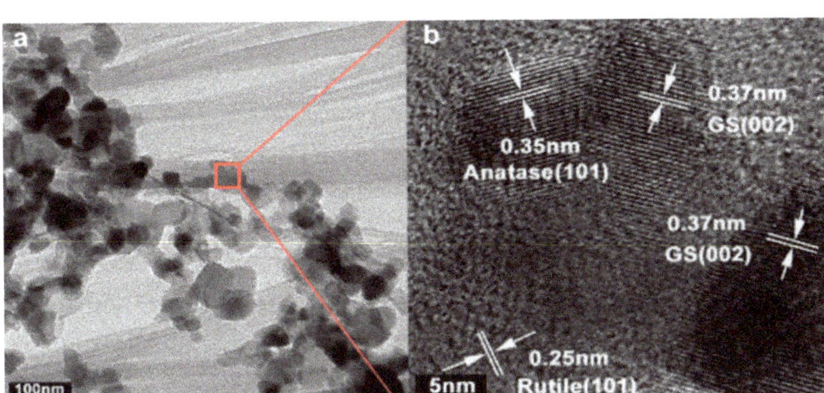

Figure 7. TEM image (**a**) and HRTEM image (**b**) of P25–10%GR [39]. (Reprinted from Int. J. Hydrogen Energy, Volume 37, Cheng P., et al., TiO_2–graphene nanocomposites for photocatalytic hydrogen production from splitting water, Pages 2224–2230. Copyright (2012), with permission from Elsevier).

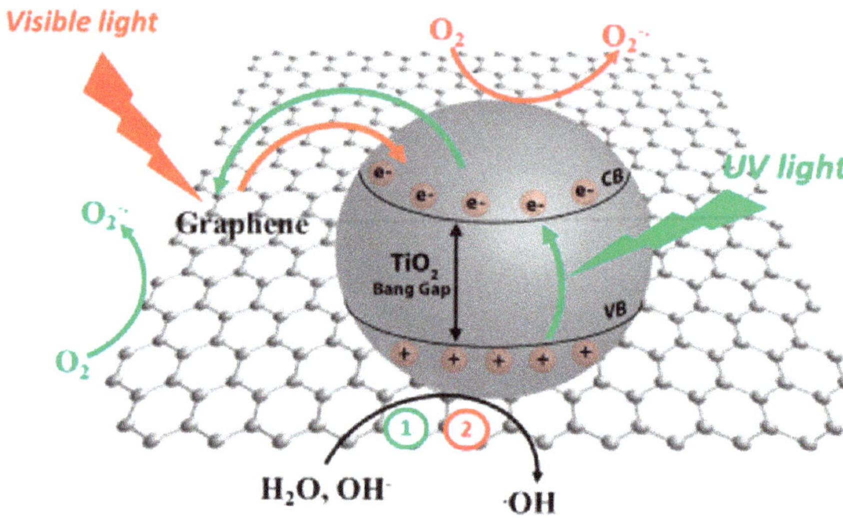

Figure 8. Mechanisms of UV and visible light activation of TiO_2 with graphene. (Reprinted from [110], Copyright (2017), MDPI).

Sankeerthana Bellamkonda et al. studied unique photocatalytic composites of graphene and carbon nanotube amalgamations of semiconductor materials. The linking of reduced TiO_2-x facets with carbon matrix exhibited the extraordinary photoactivity for the dissociation of water to produce hydrogen at room temperature. These nanohybrid composites produced hydrogen at 29,000 μmol/g·h, with 14.6% solar energy conversion efficiency, to hydrogen. The augmented photoactivity of this nanohybrid composite is ascribed to O_2 vacancies on the surface, the presence of Ti^{3+} sites in the TiO_2 lattice and the interfacial interaction between TiO_2 and graphene [114]. Multi-layers of graphene result in increased light absorption, and the formation of 111 oriented gold nanoplatelets on multilayer graphene was analyzed by Diego Mateo et al. The orientation of incorporated Au NPs was detected using XRD and EDX. The hydrogen and oxygen production rates noted were 12,000 and 9000 μmol/g·h respectively. When compared with previous literature using similar materials, it was concluded that the enhanced rates were a result of using specialized one-step synthesis process along with the highly oriented attachment of Au particles on graphene material, which affected the strength of interactions, ability of charge separation and the photo-catalytic performance [115]. Mohamed, Mokhtar et al. found that energy conversion efficiency is also prohibited in most of these metal oxides due to their high band gap energies (E_g). To mitigate such problems, spinel families, AFe_2O_4 (A = Co, Cu, Zn) have been examined as photocatalysts for H_2 production; particularly, manganese ferrite ($MnFe_2O_4$) hybridized with graphene/graphene oxide to decrease E_g and particles size, and to enhance both electronic conductivity and surface area. Enhanced light absorption ability of these hybrids induced more charge carriers, thereby increasing their apparent quantum efficiency to work as photocatalysts for water splitting. A nanospindle-shaped GO1:Mn1 hybrid absorbed more visible light for the stated purpose of generating high current density values exceeding traditional semiconductors, which improved its performance as a photocatalyst [116]. Katsuya Iwashina et al. studied the photo-catalytic activity for water splitting using the rGO-TiO_2 composite (oxygen-producing promoter) as an electron-transfer facilitator and sulfide-metal materials as hydrogen-producing promoters. Sulfide metal materials in unaided form are unfitting for direct use for water-splitting promoters due to issues relating to photo-corrosion. However, when combined with semiconducting materials for hydrogen production and using rGO-TiO_2 functionalized material for oxygen liberation, efficient gas production was observed. This scheme is also feasible for use in solar-aided reactions. Loading of noble metal co-catalysts with sulfide metals affected the activity owing to the particle sizes and extents of distribution. Moreover, the incorporation of platinum via adsorption showed the highest photo-catalytic character [117]. Zhenhua Pan et al. made RGO overcome the charge transfer challenge between loaded catalyst specks on a conductive sheet by introducing reduced graphene oxide as an arbitrating redox binder. The degree of reduction is an established factor in determining the equilibrium between hydrophilicity and conductive ability necessary for a photo-active material, which was ascertained in this case by XPS and Raman spectroscopy. The water dissociation rate was enhanced 3.5 fold by the incorporation of rGO in catalytic sheets, when compared to those that were devoid of rGO under identical circumstances, due to its inherent property of assisting in charge transmission. The role of Au as mechanical support and protection against self-oxidation was reiterated by results. Moreover, rGO helped in robust fixation of other active particulates on the Au sub-layer. Increasing its content to 1.25% (optimum) led to gradual growth in effectiveness, but a further rise in GO content caused the undesired excessive coverage of active species. Furthermore, the samples with superior dispersion of particulates presented an extra boost in efficiencies by as much as 1.7 times [118]. Graphene oxide (GO) based hybrid materials could be employed in several applications, including hydrogen generation. Carbon-based 2D structures processed by graphene oxide and its reduced form have controllable, layered structures; huge surface areas; and tunable optical, electronic and magnetic properties. Certain orthodox photocatalytic materials, such as Pt, TiO_2, ZnS and CdS, could be dispersed onto these GO/rGO matrices, resulting in efficient hybrids for less energy intensive H_2 production via photocatalytic water dissociation reaction. Doping of metal ion, particularly Co^{2+} and Al^{3+}, considerably enhanced the production of hydrogen by rGO, indicating that efficient photocatalyst could be synthesized using simple hybrids of rGO and transition metal ions [119]. A novel combination of materials was studied by Xiujun Fan, et al. Nano-crystals of three transition metal carbides were enclosed in thin graphene nano-ribbons via hot filament chemical vapor deposition technique to obtain highly pure Fe, Co and Ni carbides supported on the peaks of graphite thorn-like upright nano-ribbon structure. The stimulating hierarchical structure produced inspired a comprehensive and relative study of the sizes, shapes and other dispersion factors of carbides on 2D graphene vertical conical arrays. The GNRs not only act as a supporting substrate to enhance stability but also prevent the aggregation of carbides on the surface, holding them well into place. The porous channels also expedite the electron transport to the core, causing a distinct decline in conduction resistance. The suitability of these materials for ORR has been indicated by ample reduction of oxygen at low Tafel slopes of 39, 41 and 45 mV/dec for Fe, Co and Ni carbide GNRs, respectively—comparable to 32 mV/dec value for Pt/C electrode. The HER current densities produced by M_3C-GNRs (M = Fe, Co, Ni) were 166.6, 79.6 and 116.4 mA/cm^2 at 200 mV over-potential. These operative statistics can be ascribed to the coupled carbide-graphene effects of high roughness factors, large active area exposure and superior conducting ability and porosity of graphene [120].

2.3. Phosphorene

Phosphorene is a two dimensional (2D) material with a band gap of 0.3 to 2 eV from bulk phosphorus (Figure 9) to single layer phosphorene. The band gap of phosphorene is tunable depending on applied strain. The properties of phosphorene change with change in stress, number of layers and applied electric field [121,122]. It consists of singled-layer black phosphorus, is similar to graphene, is included in the family of monolayer-flatland materials and arranged in a hexagonal puckered lattice [50,123].

Figure 9. The layered and anisotropic crystal structure of elemental black phosphorus. (Reprinted with permission from [41]. Copyright (2015), Elsevier).

The advantage of phosphorene over graphene is that it shows a band gap. Black phosphorous was discovered around 100 years ago, but the discovery of phosphorene is just over four years old. Therefore, very little experimental research of phosphorene has been reported [124]. A monolayer of black phosphorous (BP) was synthesized using the sticky-tape technique used earlier for graphene; the material was titled phosphorene. Phosphorene has the promising properties of niche 2D materials. It shows the quantum confinement effect in a direction perpendicular to the 2D plane, which induces distinct electronic and optical properties; natural passivation of the surface without any dangling bond; no lattice mismatch concerns for assembling vertical heterostructures with other 2D materials; a high specific surface area; and a strong interaction with light. These properties are considered extremely necessary for photocatalytic applications. Additionally, phosphorene has a tunable band gap and shows anisotropic properties that make it suitable for a variety of applications [125]. Black phosphorous is the most stable allotrope of phosphorous. It has layered in-plane strong bonds along with weak van der Waals interlayer interactions. This layered structure provides the opportunity for fabrication of very thin layers of phosphorous by exfoliation from a bulk size. The techniques which have been reported for the fabrication of phosphorene so far are based on micromechanical cleavage using sticky-tape and solvent exfoliation methods. Single layer phosphorene has also been fabricated by combining mechanical exfoliation with plasma thinning technique. This sticky-tape micro mechanical exfoliation can yield phosphorene of high crystal quality, but with very low product density. In order to avoid chemical degradation, black phosphorus must be exfoliated using anhydrous and oxygen-free solvents such as acetone, chloroform, isopropyl alcohol or dimethyl form-amide. The technique of exfoliation with sonication results in low cost and a high yield of 2D phosphorene. It also seems feasible as a process for development in large scale production. However, it becomes practical only if there is a good matching of the surface tension between the layer crystal and the solvent, and if the solvent can stabilize the nanosheet to evade restacking and aggregation. Phosphorene shows a highly anisotropic structure in nature. The monolayer structure of black phosphorus is a 2D material; however, its excitonic properties closely resemble quasi-1D material, such as carbon nano-tubes. Similar to bulk black phosphorus, phosphorene shows quantum confinement perpendicular to the 2D plane. The band gap of phosphorene can be modulated appropriately for photon absorption in the ultraviolet to the near-infrared region of the solar spectrum. Apart from theoretical studies, there is actually very little published work that proves the water splitting capability of phosphorene, conclusively, at the laboratory scale [126]. However, some work has been done for phosphorene hybrids with various oxides, graphene and few sulfides. The hybrid of black phosphorous and TiO_2 has been reported as a photocatalyst in the quest for improved activity as shown in Figure 10. It was demonstrated that when bare BP was irradiated by visible light, it increased its degradation after various runs. However, when TiO_2 was loaded to form TiO_2/BP hybrid, the degradation was reduced by 90%. The reason for the improved stability can be described as due to the substitution of Ti atoms into the BP matrix to make BP resistant to moisture or oxygen. This study suggests that alteration of phosphorene with semiconductor materials might be a solution to increase the stability of phosphorene to ambient degradation and phosphorene can better be used as a photocatalyst [127]. In addition to BP/TiO_2, BP/SnO_2, BP/WO_3 and other oxides also need to be investigated, as no research has been published yet. Yongqing Cai et al. reported that graphene could be suitable as a supporting layer for encapsulating phosphorene to improve interaction and provide collaborative effects. This hybrid combination needs in depth study to find conclusive findings [128]. The eminent ability of black phosphorus nano-sheets to employ the solar energy efficiently (approximately 75%) in photocatalytic operation and the superior charge separation capacity of cobalt phosphide (CoP) can effectively be integrated in a single material, as supposed by Bin Tian, et al. [129]. The synthesis procedure consists of a straightforward solvo-thermal step involving white phosphorus and a Co-source (Co $(NO_3)_2{\cdot}6H_2O$) instead of the relatively tedious, expensive and complicated procedure of bulk BP synthesis and later exfoliation to nano-sheets. The special effects of CoP addition on BP resulted in manifold increase in H_2 release rate, 30 before and 735 μmol/g·h after CoP addition. MoS_2/phosphorene and WS_2/phosphorene hybrids demonstrate type II heterostructures. Type II is more expedient for electron-hole separation than type I and type III. The band gap of MoS_2/phosphorene varies between 0.59 to 1.06 eV, whereas it varies between 0.90 to 1.44 eV for the WS_2/phosphorene hybrid [130]. It seems that MoS_2/phosphorene cannot be used as a water splitting photocatalyst because its band gap is much below the thermodynamics band gap requisite of 1.23 eV for a photocatalyst [131]. Very little work has been done for phosphorene/sulfide hybrids. Other potential 2D sulfides/phosphorene hybrids, such as CdS/phosphorene, CdSe/phosphorene and WS_2/phosphorene, are yet to be explored. However, sulfide compounds are easily degradable in water, as they get irradiated by light. Hence, along with band gap congruency, stability is also major drawback with sulfides photocatalyst materials [132].

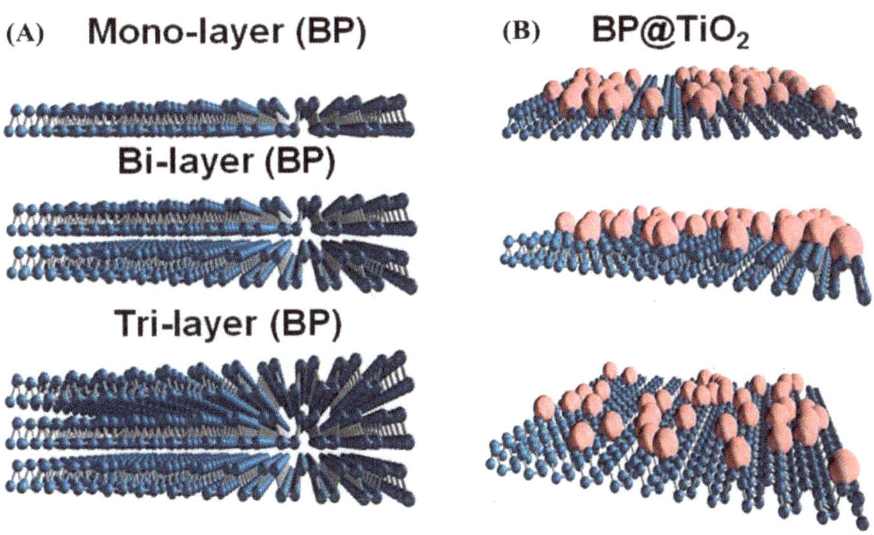

Figure 10. TiO_2 substitution on BP [127]. (Reprinted with permission from Lee, H.U., et al., Sci. Rep., 2015. 5: p. 8691).

2.4. Metal Phosphides

Phosphides of many transition metals are emerging as very popular photocatalysts due to their ability to adsorb hydrogen atoms and release H_2 gas, easily proving themselves to be very practical HER (hydrogen evolution reaction) catalysts, as confirmed by Shi and Zhang [133]. Furthermore, their tuneability to perform as dual function catalysts has been investigated via co-catalysis mechanism to enhance both HER and OER processes by incorporation of metal hydroxo/oxo groups on material surfaces. Nickel, cobalt and iron phosphides and their binary and ternary derivatives have been tried and tested by many researchers for photocatalytic applications and have showed promising outcomes. The P-content in the metal catalyst material has been suggested as a major deciding factor in determining the HER activity of material, as it is found to be responsible for the catalyst's interaction and affinity with H atoms. This trend has been reported by Yuan Pan et al. [134] and Sengeni Anantharaj, et al. [135], who confirmed the direct relation between HER catalytic activity and P content; i.e., an increase in the P ratio results in a corresponding increase in activity as well as in release rate. The OER's (oxygen evolution reaction) activity depends principally on the optimal bond energy values of the interaction between catalytic material and reaction intermediates (oxide, peroxide and hydro peroxide). Nickel phosphide was investigated as a dual functional photocatalyst by Marc Ledendecker, et al. [136]. It showed high activities for both HER and OER, attaining 10 mA/cm^2 under 1.7 V. Ni_5P_4 nanostructure was grown directly on Ni-sheets by an extremely straightforward process involving the heating of elemental phosphorus and Ni-foil sheet under inert conditions. The material showed good performance for the especially imperative OER kinetics, which is normally a challenging task in photocatalysis. This feature can be the consequence of formation of highly active NiOOH intermediate deposition on the surface which contributes to reduce the operation over-potential, the shift in electronic structure, superior current density and stable operation in a 20-h test conditions in acidic environment [136]. An alternate variation of nickel phosphide, Ni_2P as a potential electrode catalyst was suggested by Lucas A. Stern, et al. [137], wherein the active specie was found to promote both the HER and OER kinetics presenting an economically feasible and efficient single-material substitute catalyst. The Ni_2P nanoparticles and nanowires were fabricated through simple and scalable heating of the precursor ingredients, both of which showed comparable OER activities. The detailed structural studies showed the shell/core structure of NiO_x/Ni_2P which contributes prominently in the process as the Ni_2P core offers an electronic passageway to the NiO_x shell, thereby providing better diffusive characteristics and improved activity as compared to equivalent sized NiO_x nanoparticles. The reported values of current density 10 mA/cm^2 achieved with overpotential were as insignificant as 290 mV for OER and 10 mA/cm^2, with 136 mV overpotential for HER in acidic medium driving the developed catalyst material towards the mainstream photocatalytic water splitting systems as a promising and practical candidate material for both anodic and cathodic reactions [137]. Apart from nickel, other transition metal phosphides have also been investigated for catalysis to release hydrogen and oxygen from water by using incident light source. Cobalt phosphides emerged as a stimulating opportunity in this regard, and several reviewers and researchers have explored the utility of different forms and combinations of Co phosphides. Yang Y, et al. [138] formulated cobalt centric thin sheets via an intricate multistep process involving plasma cleaning of glass plate substrate, sputter-coating conduction nano-layers of chromium (10 nm) and gold (40 nm), Co-layer deposition by aqueous plating solution of $CoSO_4$, formation of porous thin Co-oxide layer by anodic treatment of Co-layer under static voltage and then finally chemical vapor deposition of phosphorus for conversion to Co-phosphide while conserving the regular porous film (500 nm thickness). The ratio of film active area to the electrode surface area (denoted by roughness factor) was calculated as a parameter to endorse the high activity of material without the necessity of external additives. Hence, the proposition was confirmed by an observed increase in RF from six (compact-before anodic treatment) to 95 (highly porous). The operational tests showed steady operation and an exceptional HER activity of 30 mA/cm^2 at overpotentials of 175 mV (in acidic condition) and 430 mV (under basic conditions) which is even higher than most noble metal based contemporary catalysts, such as CoP nanoparticles (20 mA/cm^2), Co/CNT (10 mA/cm^2), MoP (10 mA/cm^2), Ni_2P (20 mA/cm^2) and C_3N_4 (10 mA/cm^2). Similarly, notable results were observed in terms of OER activity as well; i.e., current density of 30 mA/cm^2 at 330 mV overpotential which is again higher than most up-to-date studied catalysts, such as CoP nanoparticles (25 mA/cm^2), Au/Co_3O_4 (25 mA/cm^2) and most ternary selections as well [138]. A valuable comparative analysis of different raw materials for CoP fabrication and their effectiveness in terms of HER efficiencies as scrutinized by Afriyanti Sumboja, et al. [139] to provide an outlook for large-scale and commercially applicable source of CoPs manufacture. Salts such as cobalt acetate and cobalt acetylacetonate were phosphodized by sodium hypophosphite in a single-step. The CoP formed from the former ingredient exhibited better HER promotion (overpotentials: 160 mV in acidic, 175 mV in basic media) due to greater phosphide fraction (92.5%) and greater surface area (ECSA: 10–15 cm^2). While the latter being a readily oxidizable salt produces a greater fraction of metallic Co rather than phosphide (63.5% CoP), lower surface area (ECSA: 7–12.5 cm^2) and greater overpotential (169 mV in acidic, 188 mV in basic media) [139]. The application of Co-phosphide nanosheets as HER promoter in 1M KOH media has been studied by Liang Su, et al. [140] wherein the novel approach of effects of surface refurbishment of CoP nanosheets are exposed. The CoP nanosheets were conventionally synthesized by phosphidation of Cobalt oxides and then engineered electrochemically to promote an irrevocable and stable transformation into $Co(OH)_x$/CoP hybrid configuration that displayed excellent HER catalysis producing 10 mA/cm^2 with 100 mV overpotential as compared to the untreated CoP nanosheet overpotential of 180 mV. The dispersion of hydroxyl group on the nanosheet surface as an outcome of 10 hours' exposure to 20 mA/cm^2 exhibit a coarser surface, greater turnover frequency of 0.234 per second and reliable operation even after 5000 cycles. These improvements in performance by the surface modification are attributed to the arrangement of $Co(OH)_x$ complexes which are efficient promoters of the breakdown of water into molecular components. As this stage is the rate-determining step of HER mechanism, this surface makeover of CoP nano-sheets proves to be highly favorable for the process and the copiously released H-atoms interact with adjoining CoP molecules to release hydrogen gas readily. The merger of both $Co(OH)_x$ and CoP specie is most beneficial for the process as compared to single-component systems as established by the comparative assessment of $Co(OH)_x$/CoP (100 mV overpotential), untreated CoP (180 mV) and C_3N_4 (300 mV) samples [140]. CoP nano-rods supported on Nickel foam as reported by Yun-Pie Zhu, et al. [141] is a novel strategy to fabricate dual-function HER and OER promoters with high efficacy, high porosity, enhanced diffusive transport and surface area (148 m^2/g) while maintaining high conductivity. The synthesis procedure revolves around potentio-deposition of nanorods on Ni-foam while the CoP deposited is produced in situ, facilitated by direct electron passageways provided by Ni-foam. Current densities of 10,100 and 800 mA/cm^2 at 54, 121 and 235 mV overpotentials for HER and 100 mA/cm^2 for overall cell voltage of 1.62 V were accomplished which are even more efficient than Pt and IrO_2 system [141]. Ternary variants, such as Ni-Co-P have also been developed successfully to integrate the desired properties of component materials. Yingjie Li, et al. [142] cultivated Ni-Co-P nano-sheet arrangements on Ni foam by two stage hydrothermal and phosphorization scheme. The synthesized material demonstrated extraordinary performance for both HER and OER catalysis even when compared with the traditional commercially used noble metal (Pt/C, Ir/C) electrodes and the separately tested Ni_2P and Co_2P nano-sheet arrays produced by the same method. The overpotential values for Ni-Co-P ternary electrodes for HER and OER were 133 and 308 mV respectively necessary to operate at 50 mA/cm^2 and the cell voltage required was as minimal as 1.77 V. The remarkable activity and stability of material can be credited to marginal electron resistance, normalized hydrogen binding forces, material's inherent high activity and a distinctive super-aerophobic nano-framework [142].

A similar arrangement was prepared by Liang H, et al. [143] with plasma-facilitated growth of Ni-Co-P on Ni foam, concluding in a menial over-potential of 32 mV for HER and 280 mV for OER with 10 mA/cm^2 current density and 1.58 V cell potential. Bi-functional high porosity Ni-Co-P nano-sheet catalyst supported on carbon-fiber paper was developed by Rui Wu, et al. [144] by electrodeposition resulting in identical sheet arrangements with subsequent annealing to generate spongy $Ni_{0.1}Co_{0.9}P$ structure. The annealing temperature, Ni: Co ratio of 1:9 and the precursors were deciding factors for optimal synthesis of active material. The detailed characterization results indicated that resulting material operated at high turn-over frequency (0.24 per sec for H_2, 0.12 per sec for O_2), reliable stability (20 h operation, 1000 cycles), better electron transmission ability and interactive sites contributed by Ni fraction and high efficiency. Cell voltage required for operation at 10 mA/cm^2 was a competitive value of 1.89 V and the over-potential was 125mV as compared to 174 and 196 mV for CoP and Ni_2P electrodes [144]. An ample amount of laudable work in the ternary Ni-Co-P catalyst and its composites category can be mentioned such as NiCoP/Carbon cloth composite as dual function catalyst reported by Cheng Du, et al. [145], with overpotentials as petite as 44 and 62 mV for HER and OER respectively at 10 mA/cm^2 current density and 1.77V cell potential for 100 mA/cm^2 output paves the way for practical non-noble metal electrode materials. NiCoP/r-GO composite for stimulating both HER and OER synthesized by Jiayuan Li, et al. [146] presented 10 mA/cm^2 current density with cell voltage 1.59V with stable operation for more than 75 hrs. The coupling of Co-doping in Ni_2P and their combination with rGO resulted in a boost in electron transport, catalytic activity and surface arrangement of active spots. An elegant and novel heterostructure involving NiCoP@Cu_3P nano-sheets was cultivated on Cu foam by Xingxing Ma, et al. [147] as a favorable candidate for both HER and OER catalysis showcasing 54 mV and 309 mV over-potentials for HER and OER at 10 mA/cm^2. Similarly an intricate system of NiCoP nano-sheets grown on N-doped Carbon-coated Ni foam as formulated by Miaomiao Tong, et al. [148] through hydrothermal and phosphorization route required 31.8 mV and 308.2 mV for HER and OER to accomplish 10 mA/cm^2 current density. The durable performance and activity experiments showed comparable outcomes to the commercial Pt/C and Ru_2O catalysts.

2.5. Metal Organic Frameworks and Derivatives

Metal organic framework or MOF is the peculiar category of crystalline synthetic materials that has claimed a prominent position in the league of modern multi-functional materials due to the distinctive properties of high porosity, adaptable pore structures, larger surface area and amendable elemental compositions. MOFs comprise of a network assembly of metal centers linked to organic bridging ligands which result in a variety of geometries, compositions and pore spacing dependent upon the synthesis procedure, conditions and initiators. The unique metal-centric interlinked geometric morphology allows MOF applications in various fields including, but not limited to, gas adsorption and separation, catalysis, drug delivery systems, optics and sensors [149–157] and the possibilities and opportunities of its efficient utilization and improvements are continuously being analyzed by researchers every day [158,159]. Numerous MOF-based catalyst materials have been synthesized, characterized and upgraded for photocatalytic water splitting reactions in recent times [160,161] with a focus to develop novel and feasible bi-functional promoters to make the concept of commercial solar hydrogen production from water a practical reality [162]. The standard practices to convert MOFs into practical photocatalysts comprise of inclusion of light-sensitive building blocks into the MOF backbone for instance aminoterephthalates [163,164] and porphyrins [165], Doping with any active metal or noble metal in order to enhance catalytic capacities [166], inorganic nano-particles encapsulated in MOFs [167], assembling ternary composites and introduction of electron donor or trapping materials to prevent charge recombination challenges [168]. A primary perspective of efforts in this regard will be presented herewith. The long-standing fact of high activity of noble metals in water splitting systems photocatalytically was incorporated into MOF-based materials by Meng Lan, et al. [169] to generate 3 types of Pt-tempered heterojunctions; i.e., Pt-ZnS-CoS, Pt-Zn_3P_2-CoP and Pt-ZnO-Co_3O_4. The synthesis procedure involved the ternary steps of Zn-Co ZIF preparation by previously reported methods [170–173], the sulfidation, phosphorization or oxidation of base ZnCo ZIF for each sample separately and the subsequent doping of Pt nano-particles. The twin-metal ZIF backbone facilitates the light absorption ability, increases the active area contact and favors charge carrier transportation and separation, while the dispersed Pt particles capture the electrons and motivate the hydrogen production. The resulting hydrogen release rates by above listed hetero-junctions are 8210, 9150 and 7800 µmol/g·h respectively. In the structural analysis, it was noted that the polyhedral MOF shape and porosity was retained after Pt-doping as well and the operational durability was noted to be unaffected even after 5 cycles. Cheng Wang, et al. [174] reported the fabrication of two kinds of Zr-carboxylate MOFs by intricate methods and their subsequent center-doping with Pt-nanoparticles whereby the produced Pt@MOF material displayed the ability to catalyze HER, recyclable nature by simple centrifugation and the turnover number of 7000. Another example of incorporating Pt into MOF units for enhanced activity has been reported by Dengke Wang, et al. [175] wherein, the Pt nano-particles were disseminated uniformly via a novel photo-reduction method in MIL-100(Fe) MOF which has formerly been established as a water stable and photo-excitable material [176]. The introduction of optimal fraction of Pt (0.8%) in MOF structure verified the hydrogen release increase from 5.9 to 109 μmol/g·h (18.5 times of pristine MOF). This prime value is observed due to the collaborating mechanism of electrons generation by photo-active MOF structure and the immediate electron seizing by central Pt also acting as the prime H_2 production site. This coordinated functioning results in enhanced charge separation without compromising on the exposed excitation area of for light. Qin Liu, et al. [177] used a permeable Ni-foam as a substrate to utilize its conductive capacity, high exposed surface region and ease of electronic diffusion through the channels towards the active dispersed Ni-MOF arrays which allowed the current capacity to reach 100 mA/cm^2 at 320 mV over-potential and turnover frequency of 0.25 mol/s O_2. The in-situ hydrothermally grown Ni-MOF nano-sheet arrays on Ni-foam was analyzed thoroughly and compared to commercial Ru_2O/NF OER catalyst and simple NF (Nickel foam). The results displayed the trends of lower Tafel slope 123 mV/dec than NF (193 mV/dec) but higher than Ru_2O/NF (72 mV/dec) due to the different OER mechanisms followed by these materials. Faradaic efficiency measured for oxygen release by contrasting the theoretical and practically released oxygen was nearly 100% with a stable operation for 20 h test time endorsing its practicality as a successful OER promoter. In 2016, ultra-thin nano-sheets of bi-metallic NiCo MOF were prepared with a purpose to promote oxygen discharge from water by Shenlong Zhao, et al. [178]. Utilization of nano-sized ultra-thin MOF sheets introduces the merits of lower mass transport resistance, better electronic pathways for conduction, superior exposure to the remote active sites and un-bound highly active free metal sites which act in combination with the intrinsic effectiveness of Ni and Co base metals in OER catalysis. The 2D NiCO, ultra-thin MOF nano-sheets supported on Co-sponge exhibited reassuring performance with over-potential 189 mV at 10 mA/cm^2, continuous and stable operating time of 200 h and Faradaic efficiency of 99.3%. These parameters verified the hypothesized collective benefits of incorporating two-metals in MOF network and the working decision to use ultra-thin MOF structure on a porous support. A similar study using a different combination of metals, i.e., Ni-Fe ultrathin (≈3.5 nm) MOF nano-sheets, was conducted by jingjing duan, et al. [179]. This Ni-Fe combo also showed superior analytical parameters due to the identical structural and elemental choices as previously listed work. The over-potential observed in this case was 240 mV for 10 mA/cm^2 with noteworthy turnover frequency of 3.8 per sec and stable efficient operation for 20000 s. A three-component engineered MOF structure was developed by Rui Lin, et al. [180] that combined established photo-active ingredients to make a strong partnership of UiO-66 MOF, CdS and reduced graphene oxide. Inspired to resolve the electron-hole recombination challenge in CdS photocatalysts, combining CdS/rGO composite with UiO-66 MOF not only resolved this issue but added the benefits of its high contact area, considerable thermal stability [181] uniform pore channels and proven photo-active nature of UiO-66 for hydrogen production [182]. Consequently, the UiO-66/CdS/1%-rGO combination showed the best charge separation capacity as compared to other proportions of rGO and the hydrogen release 13.8 times superior to pristine CdS. Another variation of ternary MOF photocatalyst is based on MoS_2/UiO-66/CdS fusion structure [183] which delivered 32,500 μmol/g·h H_2 release rate. A CdS/MOF combination was developed to promote H_2 evolution reaction by Jiao He, et al. [184] with the unusual idea of supporting CdS active material on photocatalytically inert MIL-101 MOF defining its sole purpose of providing greater exposed active area. MIL-101 has Cr centric structure with benefits of excellent air, water and heat stability. The supported CdS@MOF formed a crystalline cubic shape and the highest H_2 release rate was noted for 10 wt% proportion on CdS reaching upto 75,500μmol/g.h. Another specimen using MIL-101 was prepared by amine sensitization along with Pt co-catalyst doping in 2014 to obtain greater hydrogen efficiency and stability [185]. Furthermore, Co-oxide restrained in MIL-101 was prepared by intricate method involving double solvo-thermal technique to create a robust material (TOF: 0.012/sec per Co-atom) with a rich oxygen productivity (88%) [186]. The inventive strategy to produce a photo-sensitive MOF combination was explored to directly synthesize the MOF using dye-resembling organic azobenzene tetra carboxylic acid as an organic ligand and the metal (Gd) salt as the central backbone element. Xiaojun Sun, et al. [187] used a solvo-thermal method to synthesize yellow colored Gd-MOF crystals which were later also metal-laden with Ag (1–2%) as a second catalyst. The experimental comparison showed 154.2 μmol/g·h H_2 release from pristine Gd-MOF and 212 μmol/g·h from Ag (1.5%)/Gd-MOF under UV–Vis exposure due to the introduction of electron-trapping noble metal nature. Ag proved to be a suitable and feasible option as a noble metal dopant due to its cost effectiveness, lower electron transport resistance and decreased charge re-combination by capturing excited electron species released from photoactive MOF [188]. An alternate approach to integrate the qualities of superior photo-sensitivity, high HER activity and larger exposed surface area is to prepare syndicate of materials serving these purposes as formulated by Jiao He, et al. [189] wherein, a light sensitive Cr MOF UiO-66 was loaded with Pt nano-particles for HER activity and rhodamine B (RhB) dye for photo-activation. The impact of this fusion is clearly evident from H_2 evolution rates whereby pure UiO-66 displayed no perceptible H_2 release, Pt/UiO-66 showed 3.9 μmol/g·h and adsorbed RhB/Pt/UiO-66 showed 116 μmol/g·h which is 30 times greater than un-treated Pt/UiO-66. A different photo-active MOF variant MIL-125-NH_2 for H_2 production from water can be used as the base material with a MoS integration, as discovered by Tu N. Nguyen, et al. [190], to produce H_2 evolution rate as high as 2094 and 1454 μmol/g·h with quantum productivity 11% and 5.8% using two different kinds of MoS clusters; i.e., Mo_3S_{13} and IT-MoS respectively. MoS_2 is an inherently photo-active compound [191] and the sources of light responsive behavior are the vacancy defects and sulfur-edges only [192,193] while the basal faces do not contribute in catalysis and the size of particles is inversely related to the HER activity [194] Robust interaction between the supporting MOF and MoS result in higher exposed active area, lower electron conduction resistance and high stability. In an innovative approach to replicate the active nodular-structure of [NiFe] hydrogenase complex [195], a two-dimensional layered photo-responsive MOF $[Ni_2(PymS)_4]_n$ was fabricated via cautiously supervised multi-step hydrothermal process [196]. The resultant material showed robust water and air stability, bi-atomic Ni webbed morphology with 4 $PymS^-$ linkers and turnover frequency of 10.6 per hour and hydrogen release upto 6017 μmol/g·h using TEA as electron source, flourescien as photo-activator and white LED or visible light as source. The pH of solution, nature of sensitizer, nature of electron source and amount and particle size of catalyst were observed to be the deciding factors to determine the practical competence of the catalyst. A dual function catalyst with the aim of promoting both HER and OER components of the water dissociation process has many advantages and has been a center of research since decades now. A MOF-based route to achieve this goal was proposed by Yang An, et al. [169]. Pt and CoPi components were dispersed onto the innately photo-active MIL-125 (Ti) MOF [131] to boost hydrogen and oxygen discharge and suppress charge recombination tendency by the deposited components. The experimental H_2 and O_2 liberation rates under UV–Vis exposure were 42.33 and 21.33 μL/h. Separate influence of trace materials on the half-reactions were confirmed by appraising the MIL-125 (Ti) with CoPi@MIL-125 (Ti) and Pt@MIL-125 (Ti), which verified that the OER activity of CoPi-activated sample amplified to 307 μL/h from 143 μL/h, whereas in case of HER, introduction of Pt on MOF increased the activity 25 fold. Furthermore, the order of Pt and CoPi doping is a significant factor as a higher activity was noted when Pt was loaded first. Recently, an aluminum-centric MOF devised from 2-aminoterephthalic acid (ATA) was recognized as an active oxygen evolution promoter and modification of this material with Ni assimilation resulted in a dual purpose (HER and OER) overall water dissociation catalyst [197]. The half reaction analysis validated the coordinating nature of both Al and Ni components as OER for Al-ATA-Ni MOF was 5166.7 μmol/g·h contrasted with 550 μmol/g·h and HER for Al-ATA-Ni MOF was 1200 μmol/g·h and the material without Ni was found inactive towards hydrogen production. A bipyridine-centered MOF-253 was modified by confining Pt complex into the network with the intention to promote HER and simultaneously act as a photo-exciter [198]. The base MOF-253 possesses the anticipated merits of high exposed area and promising stability [199], which is evident in the resulting compound and the addition of Pt component translates to five times increase in photo-activity. Another Ti-centric MOF was amended by incorporating Ru-complex established on the similar strategy to promote HER in visible light presence [200]. Despite being a relatively new concept and material, 2D MOFs have found various application in electrocatalysis, photocatalysis and heterogeneous catalysis in general [201]. The focus on optimization of manufacturing procedure, product stability, surface coating and immobilization techniques and scalability can bring this novel category of materials into limelight for future catalytic processes due to the inherent benefits of exposed active sites, turnover frequencies, effective coordinated properties of metal and organic ligands and easy handling capability.

3. Conclusions and Future Outlook

A tremendous amount of research and development has been carried out to improve, study and implement various strategies in photocatalytic water splitting using different categories of materials, composites, heterojunctions and immobilizing schemes, as summarized in Table 2 [202,203]; however, the use of 2D materials, their combinations and derivatives is a developing concept in photocatalytic water splitting research, which poses itself as an excellent opportunity for advancement and further exploration of innovative materials. With tunable and unique architectures, distinctive features and desirable structural intricacies being the specialties of 2D materials, it seems imperative to delve into this arena for progressive photocatalysis systems. Keeping these positive outlooks and prospects in view, this review highlights the evolution of modern nanomaterials, 2D photocatalysts and MOF derivatives to attain the coveted properties of low charge recombination, high photo absorbance activity, efficient electron conductivity, fast kinetics and large surface area for water splitting to produce hydrogen. It has been concluded that production of hydrogen can be enhanced using various techniques, including doping with noble metal particles, surface functionalization, nano-heterostructures, synthesis-controlled morphology and surface decoration of nano-particles in 2D cages. The most common backbone materials noted in this regard are the graphitic carbon nitride, and graphene, which serve as supporting pillars and high area foundations in case of composite materials and hetero-structural modifications. Meanwhile, phosphorene has emerged as a relatively novel material and limited research has been done to explore the possibilities of practical application of this material. Metal phosphides and metal organic frameworks have been extensively incorporated in 2D heterostructures and as supporting cocatalysts to improve performance and functional enhancement of base materials and assure performance improvement with further research and experimentation. With the acknowledgement of the fundamentally essential properties displayed by the 2D base materials and their variants, as discussed in this article, it is safe to conclude that photocatalytic research and the commercial application of this promising technology has found 2D materials and their hybrids to be promising and stimulating candidates. However, despite the progress in research, the challenges associated with 2D materials present an opportunity to explore improvements in existing 2D materials and the development of new alternatives with novel electronic and structural features, such as electrochemically reduced thin films (e.g., α-Fe_2O_3) [28], layered oxyhalides (FeOBr, Bi_4VO_8Cl, etc.), thiophosphates ($CoPS_3$), metal chalcogenides, 2D MOFs and metal-free semiconducting materials [46]. Moreover, the setbacks associated with cost effective, bulk production of these ultrathin photocatalysts limits the scalability for prospective commercial operation which needs to be addressed and resolved. The odds of bulk scale synthesis from nano-layered and non-layered materials to manufacture 2D thin sheets must be improved to make thickness-controlled and efficient 2D photocatalysts. Debottlenecking of issues related to stability in water splitting reaction conditions, such as agglomeration, corrosion and oxidation, must be prioritized to explore surface deformation control and reduction of surface energies. Fast charge carrier re-combination could be mitigated by sacrificial agents, but better practicality demands investigation of alternative solutions, such as 2D/2D, 2D/1D and 2D/0D heterojunctions to introduce interfacial charge separation. Modern systems and industries thrive on the basis of technology; thus, optimization and simulation tools must be introduced to foster research and development in photocatalysis and eliminate the gap between theoretical and real-time reaction systems.

Table 2. Summary of photocatalytic water splitting systems discussed in this article involving g-C_3N_4, graphene, phosphine, metal phosphides and metal organic frameworks MOFs.

Photocatalyst	Bandgap (eV)	Thickness (nm)	Co-Catalyst	Sacrificial Agent	Light Source	Type of Reaction	Activity (μmol/g·h)	Ref.
g-C_3N_4 nanosheet	2.85	-	3 wt% Pt	15 vol% TEOA, $AgNO_3$ (0.1 M)	300 W Xenon lamp, λ > 420 nm cutoff filter	HER OER	297 120	[85]
Br modified g-C_3N_4	2.82	-	3 wt% Pt, 0.3 wt% CoOx	10 vol% TEOA, $AgNO_3$ (0.01 M)	300 W Xenon lamp, λ > 420 nm cutoff filter	HER OER	600 80	[86]
Ti_3C_2/g-C_3N_4 Heterojunction	2.77 of g-C_3N_4 nanosheets	1.7–2.6	3 wt% Pt	10 vol% TEOA	200 W Hg lamp, λ > 400 nm cutoff filter	HER	72.3	[89]
$NiTiO_3$/g-C_3N_4	2.18 $NiTiO_3$, 2.7 g-C_3N_4	-	1 wt% Pt	10 vol% TEOA	300 W Xenon lamp, λ > 420 nm cutoff filter	HER	835	[90]
O-doped g-C_3N_4 nanosheets	2.95	-	3 wt% Pt	-	300 W Xenon lamp, λ > 400 nm cutoff filter	HER	3786	[91]
WO3 decorated g-C_3N_4	2.39	-	-	10 vol% glycerol	500 W Xenon lamp	HER	1111	[92]
$KNbO_3$/g-C_3N_4 Heterojunctioon	-	-	2 wt% Pt	5 vol% TEOA	300 W Xenon lamp	HER	1019.38	[93]
CoO/g-C_3N_4 Heterojunctioon	2.52 CoO, 2.75 g-C_3N_4	-	-	-	White LED, λ > 400 nm	HER OER	50.2 27.8	[96]
Ni_3C/g-C_3N_4 Heterojunctioon	2.65	-	15 wt% Ni_3C	15 vol% TEOA	350 W Xenon lamp, λ > 420 nm cutoff filter	HER	1518	[97]
Ni_3N/g-C_3N_4 Heterojunctioon	-	-	3 wt% Ni_3N	10 vol% TEOA	300 W Xenon lamp, λ > 400 nm cutoff filter	HER	169	[98]
Co_2P/g-C_3N_4	2.61	-	2 wt% Co_2P	15 vol% TEOA, 0.1mM K_2HPO_4 proton carrier	300 W Xenon lamp	HER	556	[100]
ZIF-67 derived CoP/g-C_3N_4	2.85	-	1.42 wt% CoP	-	Light irradiation λ > 320 nm	HER	201.5	[101]
MnO_2/CNT/g-C_3N_4	2.64	-	MnO_2/CNT	-	70 W metal halide lamp (380 nm < λ < 780 nm)	HER	4067	[103]
GO thin film/MoS_2/g-C_3N_4 Quantum dots	1.82 MoS_2, 2.83 g-C_3N_4	-	MoS_2/g-C_3N_4 Quantum dots	0.25 M Na_2SO_3	450 W Xenon lamp	HER	1650	[104]
TiO_2/CoOx/rGO nanosheet	-	-	TiO_2 nanoparticles/CoOx	20 vol% CH_3OH, 0.05 M $AgNO_3$	250 W Hg lamp	HER OER	3800 1616	[113]
TiO_2/Graphene nanosheet	2.96	1	0.5 wt% Pt	25 vol% CH_3OH	300 W Xenon lamp	HER	6680	[39]
CNT/Graphene/TiO_2 nanohybrid	2.79	7	Graphene	10 vol% CH_3OH	350 W Xenon lamp	HER	29,000	[114]

Table 2. *Cont.*

Photocatalyst	Bandgap (eV)	Thickness (nm)	Co-Catalyst	Sacrificial Agent	Light Source	Type of Reaction	Activity (μmol/g·h)	Ref.
111-oriented Au nanoplatelets/Graphene	-	20	Au nanoplatelets	TEOA	150 W Xenon lamp	HER OER	12,000 9000	[115]
rGO-transition metal hybrids	-	-	Al, Co, Fe, Ni, Mn	-	500 W Xenon lamp	HER	24.74	[119]
Black Phosphorus/Monolayer Bi2WO6 Nanosheets	0.3–2 eV BP, 2.67 MBWO	-	3 wt% Pt	-	Visible light	HER	4208	[125]
BP/CoP nanosheets	1.14	1.4	CoP	None	300 W Xenon lamp, $\lambda > 420$ nm cutoff filter	HER	735	[129]
Pt@UiO-66-NH_2 MOF	2.76	-	2.87 wt% Pt	TEOA, CH_3CN	Visible light	HER	257.38	[155]
Pt/Amine functionalized Ti-MOF	-	-	Pt	0.01 M TEOA	Xenon lamp, $\lambda > 420$ nm	HER	366.7	[164]
Co_3O_4/TiO_2 heterojunction using Co-MOF sacrificial template	-	-	2 wt% Co	15 vol% CH_3OH	400 W Xenon lamp	HER	7000	[168]
Pt heterojunction using bimetallic MOF template	-	-	Pt	10 vol% CH_3OH	300 W Xenon lamp	HER	9150	[169]
Pt/MIL—100(Fe)	-	-	0.8 wt% Pt	CH_3OH (1:3,CH_3OH:H_2O)	300 W Xenon lamp, $\lambda > 420$ nm cutoff filter	HER	109	[175]
UiO-66/CdS/1%rGO	-	-	Pt	0.1 M Na_2S, 0.1 M Na_2SO_3	300 W Xenon lamp, $\lambda > 400$ nm cutoff filter	HER	13,800	[180]
MoS_2/UiO-66/CdS hybrid	-	-	1.5 wt% MoS_2	10 vol% LA	300 W Xenon lamp, $\lambda > 420$ nm cutoff filter	HER	32,500	[183]
CdS(10wt%)/MIL-101	-	-	0.5 wt% Pt	LA	Visible light $\lambda > 420$ nm	HER	75,500	[184]
Pt/NH_2-MIL-101(Cr)	-	-	1.5 wt% Pt	25 vol% TEOA	Visible light	HER	50,000	[185]
MoS/MIL125-NH_2	1.29 MoS_2	-	0.8 wt% $Mo_3S_{13}^{2-}$, 1T-MoS_2	16.1 vol% TEA	Xenon lamp $\lambda > 420$ nm	HER	2094, 1454	[190]
2D Ni mercaptopyrimidine MOF	-	-	-	15 vol% TEA	White LED	HER	6017	[196]
Al-based MOF derived from 2-aminoterephthalic acid	2.75	-	Ni^{2+}	CH_3OH, $AgNO_3$	Xenon lamp	HER OER	1166.7 5000	[197]

Author Contributions: All authors have contributed to the research work. Z.S. and M.U.Y. are masters students and both have written this paper under the supervision of E.P. She gave the main idea to the students about writing a comprehensive review article related to very emerging field of water splitting and guided students about how to write. M.B.K.N. have helped us a lot in grammatical purview and proof reading. All authors have read and agreed to the published version of the manuscript.

Funding: This research is funded by Higher Education Commission (HEC) Pakistan, financial grant no 2017/HEC/NRPU/10482 and APC was funded by National University of Sciences & Technology (NUST), Islamabad Pakistan.

Acknowledgments: Author Erum Pervaiz would like to acknowledge NUST Pakistan and the HEC project (2017/NRPU-10482) for financial support.

Conflicts of Interest: All authors have mutually agreed upon this submission and there is no conflict of interest that needs to be declared.

References

1. Lewis, N.S.; Nocera, D.G. Powering the planet: Chemical challenges in solar energy utilization. *Proc. Natl. Acad. Sci. USA* **2006**, *103*, 15729–15735. [CrossRef] [PubMed]
2. Zou, C.; Zhao, Q.; Zhang, G.; Xiong, B. Energy revolution: From a fossil energy era to a new energy era. *Nat. Gas Ind. B* **2016**, *3*, 1–11. [CrossRef]
3. Resch, G.; Held, A.; Faber, T.; Panzer, C.; Toro, F.; Haas, R. Potentials and prospects for renewable energies at global scale. *Energy Policy* **2008**, *36*, 4048–4056. [CrossRef]
4. Uyar, T.S.; Beşikci, D. Integration of hydrogen energy systems into renewable energy systems for better design of 100% renewable energy communities. *Int. J. Hydrogen Energy* **2017**, *42*, 2453–2456. [CrossRef]
5. Tachibana, Y.; Vayssieres, L.; Durrant, J.R. Artificial photosynthesis for solar water-splitting. *Nat. Photonics* **2012**, *6*, 511. [CrossRef]
6. Momirlan, M.; Veziroglu, T. Current status of hydrogen energy. *Renew. Sustain. Energy Rev.* **2002**, *6*, 141–179. [CrossRef]
7. Ismail, A.A.; Bahnemann, D.W. Photochemical splitting of water for hydrogen production by photocatalysis: A review. *Sol. Energy Mater. Sol. Cells* **2014**, *128*, 85–101. [CrossRef]
8. Yao, L.; Wei, D.; Ni, Y.; Yan, D.; Hu, C. Surface localization of cdzns quantum dots onto 2D g-C_3N_4 ultrathin microribbons: Highly efficient visible light-induced H_2-generation. *Nano Energy* **2016**, *26*, 248–256. [CrossRef]
9. Zhou, C.; Zhao, Y.; Shang, L.; Shi, R.; Wu, L.-Z.; Tung, C.-H.; Zhang, T. Facile synthesis of ultrathin snnb2o6 nanosheets towards improved visible-light photocatalytic h2-production activity. *Chem. Commun.* **2016**, *52*, 8239–8242. [CrossRef]
10. Inoue, Y. Photocatalytic water splitting by ruo2-loaded metal oxides and nitrides with d0- and d10 -related electronic configurations. *Energy Environ. Sci.* **2009**, *2*, 364. [CrossRef]
11. Zheng, Y.; Jiao, Y.; Jaroniec, M.; Qiao, S.Z. Advancing the electrochemistry of the hydrogen-evolution reaction through combining experiment and theory. *Angew. Chem.* **2015**, *54*, 52–65. [CrossRef] [PubMed]
12. Liang, B.; Zhang, N.; Chen, C.; Liu, X.; Ma, R.; Tong, S.; Mei, Z.; Roy, V.A.L.; Wang, H.; Tang, Y. Hierarchical yolk–shell layered potassium niobate for tuned ph-dependent photocatalytic h2 evolution. *Catal. Sci. Technol.* **2017**, *7*, 1000–1005. [CrossRef]
13. Wang, Y.; Wang, X.; Antonietti, M. Polymeric graphitic carbon nitride as a heterogeneous organocatalyst: From photochemistry to multipurpose catalysis to sustainable chemistry. *Angew. Chem.* **2012**, *51*, 68–89. [CrossRef] [PubMed]
14. Osterloh, F.E. Inorganic nanostructures for photoelectrochemical and photocatalytic water splitting. *Chem. Soc. Rev.* **2013**, *42*, 2294–2320. [CrossRef]
15. Pan, Z.; Hisatomi, T.; Wang, Q.; Chen, S.; Nakabayashi, M.; Shibata, N.; Pan, C.; Takata, T.; Katayama, M.; Minegishi, T.; et al. Photocatalyst sheets composed of particulate lamg1/3ta2/3o2n and mo-doped bivo4 for z-scheme water splitting under visible light. *ACS Catal.* **2016**, *6*, 7188–7196. [CrossRef]
16. Zhao, Y.; Zhang, J.; Li, K.; Ao, Z.; Wang, C.; Liu, H.; Sun, K.; Wang, G. Electrospun cobalt embedded porous nitrogen doped carbon nanofibers as an efficient catalyst for water splitting. *J. Mater. Chem. A* **2016**, *4*, 12818–12824. [CrossRef]
17. Hisatomi, T.; Domen, K. Reaction systems for solar hydrogen production via water splitting with particulate semiconductor photocatalysts. *Nat. Catal.* **2019**, *2*, 387–399. [CrossRef]

18. Ida, S.; Ishihara, T. Recent progress in two-dimensional oxide photocatalysts for water splitting. *J. Phys. Chem. Lett.* **2014**, *5*, 2533–2542. [CrossRef]
19. Coronado, J.M. A Historical Introduction to Photocatalysis. Design of Advanced Photocatalytic Materials for Energy and Environmental Applications. In *Green Energy and Technology*; Springer: London, UK, 2013; pp. 1–4.
20. Lee, Y.Y.; Jung, H.S.; Kang, Y.T. A review: Effect of nanostructures on photocatalytic co 2 conversion over metal oxides and compound semiconductors. *J. CO2 Util.* **2017**, *20*, 163–177. [CrossRef]
21. Wang, Q.; Hisatomi, T.; Suzuki, Y.; Pan, Z.; Seo, J.; Katayama, M.; Minegishi, T.; Nishiyama, H.; Takata, T.; Seki, K.; et al. Particulate photocatalyst sheets based on carbon conductor layer for efficient z-scheme pure-water splitting at ambient pressure. *J. Am. Chem. Soc.* **2017**, *139*, 1675–1683. [CrossRef]
22. Hisatomi, T.; Kubota, J.; Domen, K. Recent advances in semiconductors for photocatalytic and photoelectrochemical water splitting. *Chem. Soc. Rev.* **2014**, *43*, 7520–7535. [CrossRef] [PubMed]
23. Su, T.; Shao, Q.; Qin, Z.; Guo, Z.; Wu, Z. Role of interfaces in two-dimensional photocatalyst for water splitting. *ACS Catal.* **2018**, *8*, 2253–2276. [CrossRef]
24. Li, Y.; Li, Y.-L.; Sa, B.; Ahuja, R. Review of two-dimensional materials for photocatalytic water splitting from a theoretical perspective. *Catal. Sci. Technol.* **2017**, *7*, 545–559. [CrossRef]
25. Acar, C.; Dincer, I.; Naterer, G.F. Review of photocatalytic water-splitting methods for sustainable hydrogen production. *Int. J. Energy Res.* **2016**, *40*, 1449–1473. [CrossRef]
26. Tewari, V.K.; Zhang, Y. (Eds.) nanostructured two-dimensional materials. In *Modeling, Characterization, and Production of Nanomaterials*; Woodhead Publishing: Cambridge, UK, 2015; pp. 477–524.
27. Peng, R.; Ma, Y.; Huang, B.; Dai, Y. Two-dimensional janus ptsse for photocatalytic water splitting under the visible or infrared light. *J. Mater. Chem. A* **2019**, *7*, 603–610. [CrossRef]
28. Wang, J.; Waters, J.L.; Kung, P.; Kim, S.M.; Kelly, J.T.; McNamara, L.E.; Hammer, N.I.; Pemberton, B.C.; Schmehl, R.H.; Gupta, A.; et al. A facile electrochemical reduction method for improving photocatalytic performance of α-fe2o3 photoanode for solar water splitting. *ACS Appl. Mater. Interfaces* **2017**, *9*, 381–390. [CrossRef]
29. Jafari, T.; Moharreri, E.; Amin, A.S.; Miao, R.; Song, W.; Suib, S.L. Photocatalytic water splitting-the untamed dream: A review of recent advances. *Molecules* **2016**, *21*, 900. [CrossRef]
30. Murashkina, A.A.; Bakiev, T.V.; Artemev, Y.M.; Rudakova, A.V.; Emeline, A.V.; Bahnemann, D.W. Photoelectrochemical behavior of the ternary heterostructured systems cds/wo3/tio2. *Catalysts* **2019**, *9*, 999. [CrossRef]
31. Navarro Yerga, R.M.; Alvarez Galvan, M.C.; del Valle, F.; Villoria de la Mano, J.A.; Fierro, J.L. Water splitting on semiconductor catalysts under visible-light irradiation. *ChemSusChem* **2009**, *2*, 471–485. [CrossRef]
32. Shen, S.; Shi, J.; Guo, P.; Guo, L. Visible-light-driven photocatalytic water splitting on nanostructured semiconducting materials. *Int. J. Nanotechnol.* **2011**, *8*, 523–591. [CrossRef]
33. Ahmad, H.; Kamarudin, S.; Minggu, L.; Kassim, M. Hydrogen from photo-catalytic water splitting process: A review. *Renew. Sustain. Energy Rev.* **2015**, *43*, 599–610. [CrossRef]
34. Agbe, H.; Nyankson, E.; Raza, N.; Dodoo-Arhin, D.; Chauhan, A.; Osei, G.; Kumar, V.; Kim, K.-H. Recent advances in photoinduced catalysis for water splitting and environmental applications. *J. Ind. Eng. Chem.* **2019**, *72*, 31–49. [CrossRef]
35. Yousaf, M.U.; Pervaiz, E.; Minallah, S.; Afzal, M.J.; Honghong, L.; Yang, M. Tin oxide quantum dots decorated graphitic carbon nitride for enhanced removal of organic components from water: Green process. *Results Phys.* **2019**, *2019*, 102455. [CrossRef]
36. Xing, J.; Fang, W.Q.; Zhao, H.J.; Yang, H.G. Inorganic photocatalysts for overall water splitting. *Chem. Asian J.* **2012**, *7*, 642–657. [CrossRef] [PubMed]
37. Kitano, M.; Hara, M. Heterogeneous photocatalytic cleavage of water. *J. Mater. Chem.* **2010**, *20*, 627–641. [CrossRef]
38. Kouser, S.; Thannikoth, A.; Gupta, U.; Waghmare, U.V.; Rao, C.N.R. 2D-gas as a photocatalyst for water splitting to produce h2. *Small* **2015**, *11*, 4723–4730. [CrossRef]
39. Niu, P.; Zhang, L.; Liu, G.; Cheng, H.-M. Graphene-like carbon nitride nanosheets for improved photocatalytic activities. *Adv. Funct. Mater.* **2012**, *22*, 4763–4770. [CrossRef]
40. Sun, Y.; Gao, S.; Lei, F.; Xie, Y. Atomically-thin two-dimensional sheets for understanding active sites in catalysis. *Chem. Soc. Rev.* **2015**, *44*, 623–636. [CrossRef]

41. Zhuiykov, S. Nanostructured two-dimensional materials. In *Modeling, Characterization and Production of Nanomaterials: Electronics, Photonics and Energy Applications*; Woodhead Publishing: Cambridge, UK, 2015; pp. 477–524.
42. Luo, B.; Liu, G.; Wang, L. Recent advances in 2d materials for photocatalysis. *Nanoscale* **2016**, *8*, 6904–6920. [CrossRef]
43. Lu, L.; Xu, S.; Luo, Z.; Wang, S.; Li, G.; Feng, C. Synthesis of znco2o4 microspheres with zn0.33co0.67co3 precursor and their electrochemical performance. *J. Nanopart. Res.* **2016**, *18*, 183. [CrossRef]
44. Low, J.; Cao, S.; Yu, J.; Wageh, S. Two-dimensional layered composite photocatalysts. *Chem. Commun.* **2014**, *50*, 10768–10777. [CrossRef] [PubMed]
45. Ye, S.; Wang, R.; Wu, M.-Z.; Yuan, Y.-P. A review on g-c3n4 for photocatalytic water splitting and co2 reduction. *Appl. Surf. Sci.* **2015**, *358*, 15–27. [CrossRef]
46. Di, J.; Xiong, J.; Li, H.; Liu, Z. Ultrathin 2d photocatalysts: Electronic-structure tailoring, hybridization, and applications. *Adv. Mater.* **2018**, *30*, 1704548. [CrossRef] [PubMed]
47. Su, T.; Qin, Z.; Ji, H.; Wu, Z. An overview of photocatalysis facilitated by 2d heterojunctions. *Nanotechnology* **2019**, *30*, 502002. [CrossRef]
48. Zhan, W.; Sun, L.; Han, X. Recent progress on engineering highly efficient porous semiconductor photocatalysts derived from metal–organic frameworks. *Nano-Micro Lett.* **2019**, *11*, 1. [CrossRef]
49. Liu, Y.; Liang, L.; Xiao, C.; Hua, X.; Li, Z.; Pan, B.; Xie, Y. Promoting photogenerated holes utilization in pore-rich wo3 ultrathin nanosheets for efficient oxygen-evolving photoanode. *Adv. Energy Mater.* **2016**, *6*, 1600437. [CrossRef]
50. Lei, F.; Sun, Y.; Liu, K.; Gao, S.; Liang, L.; Pan, B.; Xie, Y. Oxygen vacancies confined in ultrathin indium oxide porous sheets for promoted visible-light water splitting. *J. Am. Chem. Soc.* **2014**, *136*, 6826–6829. [CrossRef]
51. Sun, Y.; Sun, Z.; Gao, S.; Cheng, H.; Liu, Q.; Lei, F.; Wei, S.; Xie, Y. All-surface-atomic-metal chalcogenide sheets for high-efficiency visible-light photoelectrochemical water splitting. *Adv. Energy Mater.* **2014**, *4*, 1300611. [CrossRef]
52. Sun, Y.; Cheng, H.; Gao, S.; Sun, Z.; Liu, Q.; Liu, Q.; Lei, F.; Yao, T.; He, J.; Wei, S.; et al. Freestanding tin disulfide single-layers realizing efficient visible-light water splitting. *Angew. Chem.* **2012**, *51*, 8727–8731. [CrossRef]
53. Sun, Y.; Sun, Z.; Gao, S.; Cheng, H.; Liu, Q.; Piao, J.; Yao, T.; Wu, C.; Hu, S.; Wei, S.; et al. Fabrication of flexible and freestanding zinc chalcogenide single layers. *Nat. Commun.* **2012**, *3*, 1057. [CrossRef]
54. Xiong, J.; Wen, L.; Jiang, F.; Liu, Y.; Liang, S.; Wu, L. Ultrathin hnb3o8 nanosheet: An efficient photocatalyst for the hydrogen production. *J. Mater. Chem. A* **2015**, *3*, 20627–20632. [CrossRef]
55. Liu, Y.; Xiong, J.; Luo, S.; Liang, R.; Qin, N.; Liang, S.; Wu, L. Ultrathin hnbwo6 nanosheets: Facile synthesis and enhanced hydrogen evolution performance from photocatalytic water splitting. *Chem. Commun.* **2015**, *51*, 15125–15128. [CrossRef] [PubMed]
56. Huang, Z.; Zhao, Y.; Song, Y.; Li, Y.; Wu, G.; Tang, H.; Zhao, J. Study on the oxidation process of cobalt hydroxide to cobalt oxides at low temperatures. *RSC Adv.* **2016**, *6*, 80059–80064. [CrossRef]
57. Xu, Y.; Zhao, W.; Xu, R.; Shi, Y.; Zhang, B. Synthesis of ultrathin cds nanosheets as efficient visible-light-driven water splitting photocatalysts for hydrogen evolution. *Chem. Commun.* **2013**, *49*, 9803–9805. [CrossRef] [PubMed]
58. Yang, W.; Zhang, L.; Xie, J.; Zhang, X.; Liu, Q.; Yao, T.; Wei, S.; Zhang, Q.; Xie, Y. Enhanced photoexcited carrier separation in oxygen-doped znin2s4 nanosheets for hydrogen evolution. *Angew. Chem. Int. Ed.* **2016**, *55*, 6716–6720. [CrossRef] [PubMed]
59. Yang, S.; Gong, Y.; Zhang, J.; Zhan, L.; Ma, L.; Fang, Z.; Vajtai, R.; Wang, X.; Ajayan, P.M. Exfoliated graphitic carbon nitride nanosheets as efficient catalysts for hydrogen evolution under visible light. *Adv. Mater.* **2013**, *25*, 2452–2456. [CrossRef]
60. Li, J.; Zhan, G.; Yu, Y.; Zhang, L. Superior visible light hydrogen evolution of janus bilayer junctions via atomic-level charge flow steering. *Nat. Commun.* **2016**, *7*, 11480. [CrossRef]
61. Okamoto, Y.; Ida, S.; Hyodo, J.; Hagiwara, H.; Ishihara, T. Synthesis and photocatalytic activity of rhodium-doped calcium niobate nanosheets for hydrogen production from a water/methanol system without cocatalyst loading. *J. Am. Chem. Soc.* **2011**, *133*, 18034–18037. [CrossRef]
62. Huang, C.; Chen, C.; Zhang, M.; Lin, L.; Ye, X.; Lin, S.; Antonietti, M.; Wang, X. Carbon-doped bn nanosheets for metal-free photoredox catalysis. *Nat. Commun.* **2015**, *6*, 7698. [CrossRef]

63. Ida, S.; Kim, N.; Ertekin, E.; Takenaka, S.; Ishihara, T. Photocatalytic reaction centers in two-dimensional titanium oxide crystals. *J. Am. Chem. Soc.* **2015**, *137*, 239–244. [CrossRef]
64. Yang, M.Q.; Xu, Y.J.; Lu, W.; Zeng, K.; Zhu, H.; Xu, Q.H.; Ho, G.W. Self-surface charge exfoliation and electrostatically coordinated 2d hetero-layered hybrids. *Nat. Commun.* **2017**, *8*, 14224. [CrossRef] [PubMed]
65. Su, T.; Hood, Z.D.; Naguib, M.; Bai, L.; Luo, S.; Rouleau, C.M.; Ivanov, I.N.; Ji, H.; Qin, Z.; Wu, Z. 2d/2d heterojunction of ti3c2/g-c3n4 nanosheets for enhanced photocatalytic hydrogen evolution. *Nanoscale* **2019**, *11*, 8138–8149. [CrossRef] [PubMed]
66. Xu, Q.; Zhu, B.; Jiang, C.; Cheng, B.; Yu, J. Constructing 2d/2d fe2o3/g-c3n4 direct z-scheme photocatalysts with enhanced h2 generation performance. *Solar RRL* **2018**, *2*, 1800006. [CrossRef]
67. Tan, P.; Zhu, A.; Qiao, L.; Zeng, W.; Ma, Y.; Dong, H.; Xie, J.; Pan, J. Constructing a direct z-scheme photocatalytic system based on 2d/2d wo3/znin2s4 nanocomposite for efficient hydrogen evolution under visible light. *Inorg. Chem. Front.* **2019**, *6*, 929–939. [CrossRef]
68. Shi, J.; Li, S.; Wang, F.; Gao, L.; Li, Y.; Zhang, X.; Lu, J. In situ topotactic formation of 2d/2d direct z-scheme cu2s/zn0.67cd0.33s in-plane intergrowth nanosheet heterojunctions for enhanced photocatalytic hydrogen production. *Dalton Trans.* **2019**, *48*, 3327–3337. [CrossRef]
69. Fu, J.; Xu, Q.; Low, J.; Jiang, C.; Yu, J. Ultrathin 2d/2d wo3/g-c3n4 step-scheme h2-production photocatalyst. *Appl. Catal. B* **2019**, *243*, 556–565. [CrossRef]
70. Cao, A.; Zhang, L.; Wang, Y.; Zhao, H.; Deng, H.; Liu, X.; Lin, Z.; Su, X.; Yue, F. 2d–2d heterostructured unimof/g-c3n4 for enhanced photocatalytic h2 production under visible-light irradiation. *ACS Sustain. Chem. Eng.* **2019**, *7*, 2492–2499. [CrossRef]
71. Yuan, Y.-J.; Shen, Z.; Wu, S.; Su, Y.; Pei, L.; Ji, Z.; Ding, M.; Bai, W.; Chen, Y.; Yu, Z.-T.; et al. Liquid exfoliation of g-c3n4 nanosheets to construct 2d-2d mos2/g-C3N4 photocatalyst for enhanced photocatalytic h2 production activity. *Appl. Catal. B* **2019**, *246*, 120–128. [CrossRef]
72. Jiang, D.; Wen, B.; Zhang, Y.; Jin, Y.; Li, D.; Chen, M. Mos2/snnb2o6 2d/2d nanosheet heterojunctions with enhanced interfacial charge separation for boosting photocatalytic hydrogen evolution. *J. Colloid Interface Sci.* **2019**, *536*, 1–8. [CrossRef]
73. Gu, W.; Li, X.; Zhang, W.; Wang, J.; Yin, X.; Zhu, L.; Chen, Z.; Zou, W.; Fu, Z.; Lu, Y. Self-limited ion-exchange grown bi6fe2ti3o18-biobr ferroelectric heterostructure and the enhanced photocatalytic oxygen evolution. *Appl. Surf. Sci.* **2019**, *479*, 137–147. [CrossRef]
74. Ning, S.; Shi, X.; Zhang, H.; Lin, H.; Zhang, Z.; Long, J.; Li, Y.; Wang, X. Reconstructing dual-induced {0 0 1} facets bismuth oxychloride nanosheets heterostructures: An effective strategy to promote photocatalytic oxygen evolution. *Solar RRL* **2019**, *3*, 1900059. [CrossRef]
75. Sun, B.; Qian, Y.; Liang, Z.; Guo, Y.; Xue, Y.; Tian, J.; Cui, H. Oxygen vacancy-rich bio2-x ultra-thin nanosheet for efficient full-spectrum responsive photocatalytic oxygen evolution from water splitting. *Sol. Energy Mater. Sol. Cells* **2019**, *195*, 309–317. [CrossRef]
76. Li, L.; She, X.; Yi, J.; Pan, L.; Xia, K.; Wei, W.; Zhu, X.; Chen, Z.; Xu, H.; Li, H. Integrating coox cocatalyst on hexagonal α-fe2o3 for effective photocatalytic oxygen evolution. *Appl. Surf. Sci.* **2019**, *469*, 933–940. [CrossRef]
77. Tang, X.; Zhao, J.-H.; Li, Y.-H.; Zhou, Z.-J.; Li, K.; Liu, F.-T.; Lan, Y.-Q. Co-doped zn1−xcdxs nanocrystals from metal–organic framework precursors: Porous microstructure and efficient photocatalytic hydrogen evolution. *Dalton Trans.* **2017**, *46*, 10553–10557. [CrossRef]
78. Zhang, M.; Huang, Y.-L.; Wang, J.-W.; Lu, T.-B. A facile method for the synthesis of a porous cobalt oxide–carbon hybrid as a highly efficient water oxidation catalyst. *J. Mater. Chem. A* **2016**, *4*, 1819–1827. [CrossRef]
79. Jiang, L.; Yuan, X.; Pan, Y.; Liang, J.; Zeng, G.; Wu, Z.; Wang, H. Doping of graphitic carbon nitride for photocatalysis: A reveiw. *Appl. Catal. B* **2017**, *217*, 388–406. [CrossRef]
80. Masih, D.; Ma, Y.; Rohani, S. Graphitic c3n4 based noble-metal-free photocatalyst systems: A review. *Appl. Catal. B* **2017**, *206*, 556–588. [CrossRef]
81. Yang, X.; Chen, Z.; Xu, J.; Tang, H.; Chen, K.; Jiang, Y. Tuning the morphology of g-c3n4 for improvement of z-scheme photocatalytic water oxidation. *ACS Appl. Mater. Interfaces* **2015**, *7*, 15285–15293. [CrossRef]
82. Cao, S.; Yu, J. G-c3n4-based photocatalysts for hydrogen generation. *J. Phys. Chem. Lett.* **2014**, *5*, 2101–2107. [CrossRef]

83. Lam, S.-M.; Sin, J.-C.; Mohamed, A.R. A review on photocatalytic application of g-c3n4/semiconductor (cns) nanocomposites towards the erasure of dyeing wastewater. *Mater. Sci. Semicond. Process.* **2016**, *47*, 62–84. [CrossRef]
84. Hu, W.; Yuan, X.; Liu, X.; Guan, Y.; Wu, X. Hierarchical sno$_2$ nanostructures as high efficient photocatalysts for the degradation of organic dyes. *J. Sol-Gel Sci. Technol.* **2017**, *84*, 316–322. [CrossRef]
85. Gu, Q.; Gao, Z.; Zhao, H.; Lou, Z.; Liao, Y.; Xue, C. Temperature-controlled morphology evolution of graphitic carbon nitride nanostructures and their photocatalytic activities under visible light. *RSC Adv.* **2015**, *5*, 49317–49325. [CrossRef]
86. Lan, Z.-A.; Zhang, G.; Wang, X. A facile synthesis of br-modified g-c3n4 semiconductors for photoredox water splitting. *Appl. Catal. B* **2016**, *192*, 116–125. [CrossRef]
87. Yan, J.; Wu, H.; Chen, H.; Zhang, Y.; Zhang, F.; Liu, S.F. Fabrication of tio2/c3n4 heterostructure for enhanced photocatalytic z-scheme overall water splitting. *Appl. Catal. B* **2016**, *191*, 130–137. [CrossRef]
88. Zhou, H.; Hu, L.; Wan, J.; Yang, R.; Yu, X.; Li, H.; Chen, J.; Wang, L.; Lu, X. Microwave-enhanced catalytic degradation of p-nitrophenol in soil using mgfe2o4. *Chem. Eng. J.* **2016**, *284*, 54–60. [CrossRef]
89. Yang, Y.; Li, X.; Lu, C.; Huang, W. G-c3n4 nanosheets coupled with tio2 nanosheets as 2d/2d heterojunction photocatalysts toward high photocatalytic activity for hydrogen production. *Catal. Lett.* **2019**, *149*, 2930–2939. [CrossRef]
90. Zeng, Y.; Wang, Y.; Chen, J.; Jiang, Y.; Kiani, M.; Li, B.; Wang, R. Fabrication of high-activity hybrid nitio3/g-c3n4 heterostructured photocatalysts for water splitting to enhanced hydrogen production. *Ceram. Int.* **2016**, *42*, 12297–12305. [CrossRef]
91. She, X.; Liu, L.; Ji, H.; Mo, Z.; Li, Y.; Huang, L.; Du, D.; Xu, H.; Li, H. Template-free synthesis of 2d porous ultrathin nonmetal-doped g-c3n4 nanosheets with highly efficient photocatalytic h2 evolution from water under visible light. *Appl. Catal. B* **2016**, *187*, 144–153. [CrossRef]
92. Kadi, M.W.; Mohamed, R.M. Increasing visible light water splitting efficiency through synthesis route and charge separation in measoporous g-c3n4 decorated with wo3 nanoparticles. *Ceram. Int.* **2019**, *45*, 3886–3893. [CrossRef]
93. Xu, D.; Li, L.; Xia, T.; Fan, W.; Wang, F.; Bai, H.; Shi, W. Heterojunction composites of g-c3n4/knbo3 enhanced photocatalytic properties for water splitting. *Int. J. Hydrogen Energy* **2018**, *43*, 16566–16572. [CrossRef]
94. Seza, A.; Soleimani, F.; Naseri, N.; Soltaninejad, M.; Montazeri, S.; Sadrnezhaad, S.; Mohammadi, M.; Moghadam, H.A.; Forouzandeh, M.; Amin, M. Novel microwave-assisted synthesis of porous g-c3n4/sno2 nanocomposite for solar water-splitting. *Appl. Surf. Sci.* **2018**, *440*, 153–161. [CrossRef]
95. Xiao, J.; Zhang, X.; Li, Y. A ternary g-c3n4/pt/zno photoanode for efficient photoelectrochemical water splitting. *Int. J. Hydrogen Energy* **2015**, *40*, 9080–9087. [CrossRef]
96. Guo, F.; Shi, W.; Zhu, C.; Li, H.; Kang, Z. Coo and g-c3n4 complement each other for highly efficient overall water splitting under visible light. *Appl. Catal. B* **2018**, *226*, 412–420. [CrossRef]
97. He, K.; Xie, J.; Liu, Z.-Q.; Li, N.; Chen, X.; Hu, J.; Li, X. Multi-functional ni3c cocatalyst/g-c3n4 nanoheterojunctions for robust photocatalytic h2 evolution under visible light. *J. Mater. Chem. A* **2018**, *6*, 13110–13122. [CrossRef]
98. Chen, L.; Huang, H.; Zheng, Y.; Sun, W.; Zhao, Y.; Francis, P.S.; Wang, X. Noble-metal-free ni3n/g-c3n4 photocatalysts with enhanced hydrogen production under visible light irradiation. *Dalton Trans.* **2018**, *47*, 12188–12196. [CrossRef]
99. Chen, S.-H.; Wang, J.-J.; Huang, J.; Li, Q.-X. G-c3n4/sns2 heterostructure: A promising water splitting photocatalyst. *Chin. J. Chem. Phys.* **2017**, *30*, 36–42. [CrossRef]
100. Shen, R.; Xie, J.; Zhang, H.; Zhang, A.; Chen, X.; Li, X. Enhanced solar fuel h2 generation over g-c3n4 nanosheet photocatalysts by the synergetic effect of noble metal-free co2p cocatalyst and the environmental phosphorylation strategy. *ACS Sustain. Chem. Eng.* **2018**, *6*, 816–826. [CrossRef]
101. Sun, X.-J.; Yang, D.-D.; Dong, H.; Meng, X.-B.; Sheng, J.-L.; Zhang, X.; Wei, J.-Z.; Zhang, F.-M. Zif-derived cop as a cocatalyst for enhanced photocatalytic h2 production activity of g-c3n4. *Sustain. Energy Fuels* **2018**, *2*, 1356–1361. [CrossRef]
102. Wang, Y.; Liu, X.; Liu, J.; Han, B.; Hu, X.; Yang, F.; Xu, Z.; Li, Y.; Jia, S.; Li, Z.; et al. Carbon quantum dot implanted graphite carbon nitride nanotubes: Excellent charge separation and enhanced photocatalytic hydrogen evolution. *Angew. Chem.* **2018**, *57*, 5765–5771. [CrossRef]

103. Wang, N.; Li, J.; Wu, L.; Li, X.; Shu, J. Mno2 and carbon nanotube co-modified c3n4 composite catalyst for enhanced water splitting activity under visible light irradiation. *Int. J. Hydrogen Energy* **2016**, *41*, 22743–22750. [CrossRef]
104. Wang, M.; Ju, P.; Li, J.; Zhao, Y.; Han, X.; Hao, Z. Facile synthesis of mos2/g-c3n4/go ternary heterojunction with enhanced photocatalytic activity for water splitting. *ACS Sustain. Chem. Eng.* **2017**, *5*, 7878–7886. [CrossRef]
105. Schedin, F.; Geim, A.; Morozov, S.; Hill, E.; Blake, P.; Katsnelson, M.; Novoselov, K. Detection of individual gas molecules adsorbed on graphene. *Nat. Mater.* **2007**, *6*, 652. [CrossRef] [PubMed]
106. Warner, J.H.; Schäffel, F.; Bachmatiuk, A.; Rümmeli, M.H. (Eds.) Chapter 3—Properties of graphene. In *Graphene, Warner*; Woodhead Publishing: Cambridge, UK, 2013; pp. 61–127.
107. Choi, W.; Lahiri, I.; Seelaboyina, R.; Kang, Y.S. Synthesis of graphene and its applications: A review. *Crit. Rev. Solid State Mater. Sci.* **2010**, *35*, 52–71. [CrossRef]
108. Montes-Navajas, P.; Asenjo, N.G.; Santamaria, R.; Menendez, R.; Corma, A.; Garcia, H. Surface area measurement of graphene oxide in aqueous solutions. *Langmuir ACS J. Surf. Colloids* **2013**, *29*, 13443–13448. [CrossRef]
109. Lv, X.-J.; Zhou, S.-X.; Zhang, C.; Chang, H.-X.; Chen, Y.; Fu, W.-F. Synergetic effect of cu and graphene as cocatalyst on tio 2 for enhanced photocatalytic hydrogen evolution from solar water splitting. *J. Mater. Chem.* **2012**, *22*, 18542–18549. [CrossRef]
110. Giovannetti, R.; Rommozzi, E.; Zannotti, M.; D'Amato, C.A. Recent advances in graphene based tio2 nanocomposites (gtio2ns) for photocatalytic degradation of synthetic dyes. *Catalysts* **2017**, *7*, 305. [CrossRef]
111. Yan, Y.; Chen, J.; Li, N.; Tian, J.; Li, K.; Jiang, J.; Liu, J.; Tian, Q.; Chen, P. Systematic bandgap engineering of graphene quantum dots and applications for photocatalytic water splitting and co2 reduction. *ACS Nano* **2018**, *12*, 3523–3532. [CrossRef]
112. Del Pino, A.P.; González-Campo, A.; Giraldo, S.; Peral, J.; György, E.; Logofatu, C.; Puigmartí-Luis, J. Synthesis of graphene-based photocatalysts for water splitting by laser-induced doping with ionic liquids. *Carbon* **2018**, *130*, 48–58. [CrossRef]
113. Min, S.; Wang, F.; Lu, G. Graphene-induced spatial charge separation for selective water splitting over tio2 photocatalyst. *Catal. Commun.* **2016**, *80*, 28–32. [CrossRef]
114. Bellamkonda, S.; Thangavel, N.; Hafeez, H.Y.; Neppolian, B.; Rao, G.R. Highly active and stable multi-walled carbon nanotubes-graphene-tio2 nanohybrid: An efficient non-noble metal photocatalyst for water splitting. *Catal. Today* **2019**, *321*, 120–127. [CrossRef]
115. Mateo, D.; Esteve-Adell, I.; Albero, J.; Royo, J.F.S.; Primo, A.; Garcia, H. 111 oriented gold nanoplatelets on multilayer graphene as visible light photocatalyst for overall water splitting. *Nat. Commun.* **2016**, *7*, 11819. [CrossRef] [PubMed]
116. Mohamed, M.M.; Ibrahim, I.; Salama, T.M. Rational design of manganese ferrite-graphene hybrid photocatalysts: Efficient water splitting and effective elimination of organic pollutants. *Appl. Catal. A* **2016**, *524*, 182–191. [CrossRef]
117. Iwashina, K.; Iwase, A.; Ng, Y.H.; Amal, R.; Kudo, A. Z-schematic water splitting into h2 and o2 using metal sulfide as a hydrogen-evolving photocatalyst and reduced graphene oxide as a solid-state electron mediator. *J. Am. Chem. Soc.* **2015**, *137*, 604–607. [CrossRef] [PubMed]
118. Pan, Z.; Hisatomi, T.; Wang, Q.; Chen, S.; Iwase, A.; Nakabayashi, M.; Shibata, N.; Takata, T.; Katayama, M.; Minegishi, T. Photoreduced graphene oxide as a conductive binder to improve the water splitting activity of photocatalyst sheets. *Adv. Funct. Mater.* **2016**, *26*, 7011–7019. [CrossRef]
119. Shudo, Y.; Karim, M.R.; Wakata, K.; Ohmagari, H.; Kameda, N.; Hayami, S. Reduced graphene oxide-transition metal hybrids for hydrogen generation by photocatalytic water splitting. *J. Incl. Phenom. Macrocycl. Chem.* **2018**, 1–4. [CrossRef]
120. Fan, X.; Peng, Z.; Ye, R.; Zhou, H.; Guo, X. M3c (m: Fe, co, ni) nanocrystals encased in graphene nanoribbons: An active and stable bifunctional electrocatalyst for oxygen reduction and hydrogen evolution reactions. *ACS Nano* **2015**, *9*, 7407–7418. [CrossRef]
121. Jing, Y.; Tang, Q.; He, P.; Zhou, Z.; Shen, P. Small molecules make big differences: Molecular doping effects on electronic and optical properties of phosphorene. *Nanotechnology* **2015**, *26*, 095201. [CrossRef]
122. Khandelwal, A.; Mani, K.; Karigerasi, M.H.; Lahiri, I. Phosphorene–the two-dimensional black phosphorous: Properties, synthesis and applications. *Mater. Sci. Eng. B* **2017**, *221*, 17–34. [CrossRef]

123. Churchill, H.O.; Jarillo-Herrero, P. Two-dimensional crystals: Phosphorus joins the family. *Nat. Nanotechnol.* **2014**, *9*, 330. [CrossRef]
124. Lee, T.H.; Kim, S.Y.; Jang, H.W. Black phosphorus: Critical review and potential for water splitting photocatalyst. *Nanomaterials* **2016**, *6*, 194. [CrossRef]
125. Hu, J.; Chen, D.; Mo, Z.; Li, N.; Xu, Q.; Li, H.; He, J.; Xu, H.; Lu, J. Z-scheme 2d/2d heterojunction of black phosphorus/monolayer bi2wo6 nanosheets with enhanced photocatalytic activities. *Angew. Chem. Int. Ed.* **2019**, *58*, 2073–2077. [CrossRef] [PubMed]
126. Kou, L.; Chen, C.; Smith, S.C. Phosphorene: Fabrication, properties, and applications. *J. Phys. Chem. Lett.* **2015**, *6*, 2794–2805. [CrossRef] [PubMed]
127. Lee, H.U.; Lee, S.C.; Won, J.; Son, B.-C.; Choi, S.; Kim, Y.; Park, S.Y.; Kim, H.-S.; Lee, Y.-C.; Lee, J. Stable semiconductor black phosphorus (bp)@ titanium dioxide (tio 2) hybrid photocatalysts. *Sci. Rep.* **2015**, *5*, 8691. [CrossRef] [PubMed]
128. Cai, Y.; Zhang, G.; Zhang, Y.-W. Electronic properties of phosphorene/graphene and phosphorene/hexagonal boron nitride heterostructures. *J. Phys. Chem. C* **2015**, *119*, 13929–13936. [CrossRef]
129. Tian, B.; Tian, B.; Smith, B.; Scott, M.C.; Hua, R.; Lei, Q.; Tian, Y. Supported black phosphorus nanosheets as hydrogen-evolving photocatalyst achieving 5.4% energy conversion efficiency at 353 k. *Nat. Commun.* **2018**, *9*, 1397. [CrossRef]
130. Guo, H.; Lu, N.; Dai, J.; Wu, X.; Zeng, X.C. Phosphorene nanoribbons, phosphorus nanotubes, and van der waals multilayers. *J. Phys. Chem. C* **2014**, *118*, 14051–14059. [CrossRef]
131. Huang, L.; Huo, N.; Li, Y.; Chen, H.; Yang, J.; Wei, Z.; Li, J.; Li, S.-S. Electric-field tunable band offsets in black phosphorus and mos2 van der waals pn heterostructure. *J. Phys. Chem. Lett.* **2015**, *6*, 2483–2488. [CrossRef]
132. Yuan, Y.-P.; Ruan, L.-W.; Barber, J.; Loo, S.C.J.; Xue, C. Hetero-nanostructured suspended photocatalysts for solar-to-fuel conversion. *Energy Environ. Sci.* **2014**, *7*, 3934–3951. [CrossRef]
133. Shi, Y.; Zhang, B. Recent advances in transition metal phosphide nanomaterials: Synthesis and applications in hydrogen evolution reaction. *Chem. Soc. Rev.* **2016**, *45*, 1529–1541. [CrossRef]
134. Pan, Y.; Liu, Y.; Zhao, J.; Yang, K.; Liang, J.; Liu, D.; Hu, W.; Liu, D.; Liu, Y.; Liu, C. Monodispersed nickel phosphide nanocrystals with different phases: Synthesis, characterization and electrocatalytic properties for hydrogen evolution. *J. Mater. Chem. A* **2015**, *3*, 1656–1665. [CrossRef]
135. Anantharaj, S.; Ede, S.R.; Sakthikumar, K.; Karthick, K.; Mishra, S.; Kundu, S. Recent trends and perspectives in electrochemical water splitting with an emphasis on sulfide, selenide, and phosphide catalysts of fe, co, and ni: A review. *ACS Catal.* **2016**, *6*, 8069–8097. [CrossRef]
136. Ledendecker, M.; Krick Calderón, S.; Papp, C.; Steinrück, H.-P.; Antonietti, M.; Shalom, M. The synthesis of nanostructured ni5p4 films and their use as a non-noble bifunctional electrocatalyst for full water splitting. *Angew. Chem. Int. Ed.* **2015**, *54*, 12361–12365. [CrossRef] [PubMed]
137. Stern, L.-A.; Feng, L.; Song, F.; Hu, X. Ni2p as a janus catalyst for water splitting: The oxygen evolution activity of ni2p nanoparticles. *Energy Environ. Sci.* **2015**, *8*, 2347–2351. [CrossRef]
138. Yang, Y.; Fei, H.; Ruan, G.; Tour, J.M. Porous cobalt-based thin film as a bifunctional catalyst for hydrogen generation and oxygen generation. *Adv. Mater.* **2015**, *27*, 3175–3180. [CrossRef] [PubMed]
139. Sumboja, A.; An, T.; Goh, H.Y.; Lubke, M.; Howard, D.P.; Xu, Y.; Handoko, A.D.; Zong, Y.; Liu, Z. One-step facile synthesis of cobalt phosphides for hydrogen evolution reaction catalysts in acidic and alkaline medium. *ACS Appl. Mater. Interfaces* **2018**, *10*, 15673–15680. [CrossRef] [PubMed]
140. Su, L.; Cui, X.; He, T.; Zeng, L.; Tian, H.; Song, Y.; Qi, K.; Xia, B.Y. Surface reconstruction of cobalt phosphide nanosheets by electrochemical activation for enhanced hydrogen evolution in alkaline solution. *Chem. Sci.* **2019**, *10*, 2019–2024. [CrossRef]
141. Zhu, Y.-P.; Liu, Y.-P.; Ren, T.-Z.; Yuan, Z.-Y. Self-supported cobalt phosphide mesoporous nanorod arrays: A flexible and bifunctional electrode for highly active electrocatalytic water reduction and oxidation. *Adv. Funct. Mater.* **2015**, *25*, 7337–7347. [CrossRef]
142. Li, Y.; Zhang, H.; Jiang, M.; Kuang, Y.; Sun, X.; Duan, X. Ternary nicop nanosheet arrays: An excellent bifunctional catalyst for alkaline overall water splitting. *Nano Res.* **2016**, *9*, 2251–2259. [CrossRef]
143. Liang, H.; Gandi, A.N.; Anjum, D.H.; Wang, X.; Schwingenschlogl, U.; Alshareef, H.N. Plasma-assisted synthesis of nicop for efficient overall water splitting. *Nano Lett.* **2016**, *16*, 7718–7725. [CrossRef]

144. Wu, R.; Xiao, B.; Gao, Q.; Zheng, Y.-R.; Zheng, X.-S.; Zhu, J.-F.; Gao, M.-R.; Yu, S.-H. A janus nickel cobalt phosphide catalyst for high-efficiency neutral-ph water splitting. *Angew. Chem. Int. Ed.* **2018**, *57*, 15445–15449. [CrossRef]
145. Du, C.; Yang, L.; Yang, F.; Cheng, G.; Luo, W. Nest-like nicop for highly efficient overall water splitting. *ACS Catal.* **2017**, *7*, 4131–4137. [CrossRef]
146. Orlandi, M.; Brenna, D.; Harms, R.; Jost, S.; Benaglia, M. Recent developments in the reduction of aromatic and aliphatic nitro compounds to amines. *Org. Process Res. Dev.* **2016**, *22*, 430–445. [CrossRef]
147. Ma, X.; Chang, Y.; Zhang, Z.; Tang, J. Forest-like nicop@cu3p supported on copper foam as a bifunctional catalyst for efficient water splitting. *J. Mater. Chem. A* **2018**, *6*, 2100–2106. [CrossRef]
148. Tong, M.; Wang, L.; Yu, P.; Liu, X.; Fu, H. 3d network nanostructured nicop nanosheets supported on n-doped carbon coated ni foam as a highly active bifunctional electrocatalyst for hydrogen and oxygen evolution reactions. *Front. Chem. Sci. Eng.* **2018**, *12*, 417–424. [CrossRef]
149. Du, D.Y.; Qin, J.S.; Li, S.L.; Su, Z.M.; Lan, Y.Q. Recent advances in porous polyoxometalate-based metal-organic framework materials. *Chem. Soc. Rev.* **2014**, *43*, 4615–4632. [CrossRef] [PubMed]
150. Bordiga, S.; Lamberti, C.; Ricchiardi, G.; Regli, L.; Bonino, F.; Damin, A.; Lillerud, K.P.; Bjorgen, M.; Zecchina, A. Electronic and vibrational properties of a mof-5 metal-organic framework: Zno quantum dot behaviour. *Chem. Commun. (Camb. Engl.)* **2004**, 2300–2301. [CrossRef]
151. Tachikawa, T.; Choi, J.R.; Fujitsuka, M.; Majima, T. Photoinduced charge-transfer processes on mof-5 nanoparticles: Elucidating differences between metal-organic frameworks and semiconductor metal oxides. *J. Phys. Chem. C* **2008**, *112*, 14090–14101. [CrossRef]
152. Bauer, C.A.; Timofeeva, T.V.; Settersten, T.B.; Patterson, B.D.; Liu, V.H.; Simmons, B.A.; Allendorf, M.D. Influence of connectivity and porosity on ligand-based luminescence in zinc metal-organic frameworks. *J. Am. Chem. Soc.* **2007**, *129*, 7136–7144. [CrossRef]
153. Li, F.; Jiang, X.; Zhao, J.; Zhang, S. Graphene oxide: A promising nanomaterial for energy and environmental applications. *Nano Energy* **2015**, *16*, 488–515. [CrossRef]
154. Zhang, X.; Liu, S.; Zang, Y.; Liu, R.; Liu, G.; Wang, G.; Zhang, Y.; Zhang, H.; Zhao, H. Co/co 9 s 8 @s,n-doped porous graphene sheets derived from s, n dual organic ligands assembled co-mofs as superior electrocatalysts for full water splitting in alkaline media. *Nano Energy* **2016**, *30*, 93–102. [CrossRef]
155. Xiao, J.-D.; Shang, Q.; Xiong, Y.; Zhang, Q.; Luo, Y.; Yu, S.-H.; Jiang, H.-L. Boosting photocatalytic hydrogen production of a metal–organic framework decorated with platinum nanoparticles: The platinum location matters. *Angew. Chem.* **2016**, *128*, 9535–9539. [CrossRef]
156. Qin, J.S.; Zhang, S.R.; Du, D.Y.; Shen, P.; Bao, S.J.; Lan, Y.Q.; Su, Z.M. A microporous anionic metal-organic framework for sensing luminescence of lanthanide(iii) ions and selective absorption of dyes by ionic exchange. *Chemistry* **2014**, *20*, 5625–5630. [CrossRef] [PubMed]
157. Gao, C.Y.; Tian, H.R.; Ai, J.; Li, L.J.; Dang, S.; Lan, Y.Q.; Sun, Z.M. A microporous cu-mof with optimized open metal sites and pore spaces for high gas storage and active chemical fixation of co2. *Chem. Commun. (Camb. Engl.)* **2016**, *52*, 11147–11150. [CrossRef] [PubMed]
158. Gascon, J.; Corma, A.; Kapteijn, F.; Llabrés i Xamena, F.X. Metal organic framework catalysis: Quo vadis? *ACS Catal.* **2014**, *4*, 361–378. [CrossRef]
159. Wang, C.; Liu, D.; Lin, W. Metal-organic frameworks as a tunable platform for designing functional molecular materials. *J. Am. Chem. Soc.* **2013**, *135*, 13222–13234. [CrossRef]
160. Zhang, T.; Lin, W. Metal-organic frameworks for artificial photosynthesis and photocatalysis. *Chem. Soc. Rev.* **2014**, *43*, 5982–5993. [CrossRef]
161. Song, F.; Li, W.; Sun, Y. Metal–organic frameworks and their derivatives for photocatalytic water splitting. *Inorganics* **2017**, *5*, 40. [CrossRef]
162. Wang, W.; Xu, X.; Zhou, W.; Shao, Z. Recent progress in metal-organic frameworks for applications in electrocatalytic and photocatalytic water splitting. *Adv. Sci.* **2017**, *4*, 1600371. [CrossRef]
163. Long, J.; Wang, S.; Ding, Z.; Wang, S.; Zhou, Y.; Huang, L.; Wang, X. Amine-functionalized zirconium metal-organic framework as efficient visible-light photocatalyst for aerobic organic transformations. *Chem. Commun. (Camb. Engl.)* **2012**, *48*, 11656–11658. [CrossRef]
164. Horiuchi, Y.; Toyao, T.; Saito, M.; Mochizuki, K.; Iwata, M.; Higashimura, H.; Anpo, M.; Matsuoka, M. Visible-light-promoted photocatalytic hydrogen production by using an amino-functionalized ti(iv) metal–organic framework. *J. Phys. Chem. C* **2012**, *116*, 20848–20853. [CrossRef]

165. Fateeva, A.; Chater, P.A.; Ireland, C.P.; Tahir, A.A.; Khimyak, Y.Z.; Wiper, P.V.; Darwent, J.R.; Rosseinsky, M.J. A water-stable porphyrin-based metal-organic framework active for visible-light photocatalysis. *Angew. Chem.* **2012**, *51*, 7440–7444. [CrossRef] [PubMed]
166. Wang, C.; Xie, Z.; deKrafft, K.E.; Lin, W. Doping metal-organic frameworks for water oxidation, carbon dioxide reduction, and organic photocatalysis. *J. Am. Chem. Soc.* **2011**, *133*, 13445–13454. [CrossRef] [PubMed]
167. Xiang, W.; Zhang, Y.; Lin, H.; Liu, C.J. Nanoparticle/metal-organic framework composites for catalytic applications: Current status and perspective. *Molecules* **2017**, *22*, 2103. [CrossRef] [PubMed]
168. Bala, S.; Mondal, I.; Goswami, A.; Pal, U.; Mondal, R. Co-mof as a sacrificial template: Manifesting a new co3o4/tio2 system with a p–n heterojunction for photocatalytic hydrogen evolution. *J. Mater. Chem. A* **2015**, *3*, 20288–20296. [CrossRef]
169. Lan, M.; Guo, R.-M.; Dou, Y.; Zhou, J.; Zhou, A.; Li, J.-R. Fabrication of porous pt-doping heterojunctions by using bimetallic mof template for photocatalytic hydrogen generation. *Nano Energy* **2017**, *33*, 238–246. [CrossRef]
170. Jiang, Z.; Lu, W.; Li, Z.; Ho, K.H.; Li, X.; Jiao, X.; Chen, D. Synthesis of amorphous cobalt sulfide polyhedral nanocages for high performance supercapacitors. *J. Mater. Chem. A* **2014**, *2*, 8603–8606. [CrossRef]
171. Venna, S.R.; Jasinski, J.B.; Carreon, M.A. Structural evolution of zeolitic imidazolate framework-8. *J. Am. Chem. Soc.* **2010**, *132*, 18030–18033. [CrossRef]
172. Yu, B.; Zhang, D.; Du, S.; Wang, Y.; Chen, M.; Hou, J.; Xu, S.; Wu, S.; Gong, J. Phase transfer directed synthesis of hollow zeolitic imidazolate frameworks-67 nanocages. *Cryst. Growth Des.* **2016**, *17*, 3–6. [CrossRef]
173. Xu, W.; Li, T.-T.; Zheng, Y.-Q. Porous co3o4 nanoparticles derived from a co(ii)-cyclohexanehexacarboxylate metal–organic framework and used in a supercapacitor with good cycling stability. *RSC Adv.* **2016**, *6*, 86447–86454. [CrossRef]
174. Wang, C.; deKrafft, K.E.; Lin, W. Pt nanoparticles@photoactive metal–organic frameworks: Efficient hydrogen evolution via synergistic photoexcitation and electron injection. *J. Am. Chem. Soc.* **2012**, *134*, 7211–7214. [CrossRef]
175. Wang, D.; Song, Y.; Cai, J.; Wu, L.; Li, Z. Effective photo-reduction to deposit pt nanoparticles on mil-100(fe) for visible-light-induced hydrogen evolution. *New J. Chem.* **2016**, *40*, 9170–9175. [CrossRef]
176. Laurier, K.G.; Vermoortele, F.; Ameloot, R.; De Vos, D.E.; Hofkens, J.; Roeffaers, M.B. Iron(iii)-based metal-organic frameworks as visible light photocatalysts. *J. Am. Chem. Soc.* **2013**, *135*, 14488–14491. [CrossRef]
177. Liu, Q.; Xie, L.; Shi, X.; Du, G.; Asiri, A.M.; Luo, Y.; Sun, X. High-performance water oxidation electrocatalysis enabled by a ni-mof nanosheet array. *Inorg. Chem. Front.* **2018**, *5*, 1570–1574. [CrossRef]
178. Zhao, S.; Wang, Y.; Dong, J.; He, C.-T.; Yin, H.; An, P.; Zhao, K.; Zhang, X.; Gao, C.; Zhang, L.; et al. Ultrathin metal–organic framework nanosheets for electrocatalytic oxygen evolution. *Nat. Energy* **2016**, *1*, 16184. [CrossRef]
179. Duan, J.; Chen, S.; Zhao, C. Ultrathin metal-organic framework array for efficient electrocatalytic water splitting. *Nat. Commun.* **2017**, *8*, 15341. [CrossRef]
180. Lin, R.; Shen, L.; Ren, Z.; Wu, W.; Tan, Y.; Fu, H.; Zhang, J.; Wu, L. Enhanced photocatalytic hydrogen production activity via dual modification of mof and reduced graphene oxide on cds. *Chem. Commun.* **2014**, *50*, 8533–8535. [CrossRef]
181. Cavka, J.H.; Jakobsen, S.; Olsbye, U.; Guillou, N.; Lamberti, C.; Bordiga, S.; Lillerud, K.P. A new zirconium inorganic building brick forming metal organic frameworks with exceptional stability. *J. Am. Chem. Soc.* **2008**, *130*, 13850–13851. [CrossRef]
182. Gomes Silva, C.; Luz, I.; Llabres i Xamena, F.X.; Corma, A.; Garcia, H. Water stable zr-benzenedicarboxylate metal-organic frameworks as photocatalysts for hydrogen generation. *Chemistry* **2010**, *16*, 11133–11138. [CrossRef]
183. Shen, L.; Luo, M.; Liu, Y.; Liang, R.; Jing, F.; Wu, L. Noble-metal-free mos2 co-catalyst decorated uio-66/cds hybrids for efficient photocatalytic h2 production. *Appl. Catal. B* **2015**, *166–167*, 445–453. [CrossRef]
184. He, J.; Yan, Z.; Wang, J.; Xie, J.; Jiang, L.; Shi, Y.; Yuan, F.; Yu, F.; Sun, Y. Significantly enhanced photocatalytic hydrogen evolution under visible light over cds embedded on metal–organic frameworks. *Chem. Commun.* **2013**, *49*, 6761–6763. [CrossRef]

185. Wen, M.; Mori, K.; Kamegawa, T.; Yamashita, H. Amine-functionalized mil-101(cr) with imbedded platinum nanoparticles as a durable photocatalyst for hydrogen production from water. *Chem. Commun.* **2014**, *50*, 11645–11648. [CrossRef]
186. Han, J.; Wang, D.; Du, Y.; Xi, S.; Hong, J.; Yin, S.; Chen, Z.; Zhou, T.; Xu, R. Metal–organic framework immobilized cobalt oxide nanoparticles for efficient photocatalytic water oxidation. *J. Mater. Chem. A* **2015**, *3*, 20607–20613. [CrossRef]
187. Sun, X.; Yu, Q.; Zhang, F.; Wei, J.; Yang, P. A dye-like ligand-based metal–organic framework for efficient photocatalytic hydrogen production from aqueous solution. *Catal. Sci. Technol.* **2016**, *6*, 3840–3844. [CrossRef]
188. Yang, D.; Sun, Y.; Tong, Z.; Tian, Y.; Li, Y.; Jiang, Z. Synthesis of ag/tio2 nanotube heterojunction with improved visible-light photocatalytic performance inspired by bioadhesion. *J. Phys. Chem. C* **2015**, *119*, 5827–5835. [CrossRef]
189. He, J.; Wang, J.; Chen, Y.; Zhang, J.; Duan, D.; Wang, Y.; Yan, Z. A dye-sensitized pt@uio-66(zr) metal-organic framework for visible-light photocatalytic hydrogen production. *Chem. Commun. (Camb. Engl.)* **2014**, *50*, 7063–7066. [CrossRef]
190. Nguyen, T.N.; Kampouri, S.; Valizadeh, B.; Luo, W.; Ongari, D.; Planes, O.M.; Zuttel, A.; Smit, B.; Stylianou, K.C. Photocatalytic hydrogen generation from a visible-light-responsive metal-organic framework system: Stability versus activity of molybdenum sulfide cocatalysts. *ACS Appl. Mater. Interfaces* **2018**, *10*, 30035–30039. [CrossRef]
191. Rao, C.N.R.; Maitra, U.; Waghmare, U.V. Extraordinary attributes of 2-dimensional mos2 nanosheets. *Chem. Phys. Lett.* **2014**, *609*, 172–183. [CrossRef]
192. Jaramillo, T.F.; Jorgensen, K.P.; Bonde, J.; Nielsen, J.H.; Horch, S.; Chorkendorff, I. Identification of active edge sites for electrochemical h2 evolution from mos2 nanocatalysts. *Science (N. Y.)* **2007**, *317*, 100–102. [CrossRef]
193. Ataca, C.; Ciraci, S. Dissociation of h2o at the vacancies of single-layer mos2. *Biophys. Rev. B* **2012**, *85*, 195410. [CrossRef]
194. Laursen, A.B.; Kegnæs, S.; Dahl, S.; Chorkendorff, I. Molybdenum sulfides—efficient and viable materials for electro - and photoelectrocatalytic hydrogen evolution. *Energy Environ. Sci.* **2012**, *5*, 5577. [CrossRef]
195. Van Gastel, M.; Shaw, J.L.; Blake, A.J.; Flores, M.; Schröder, M.; McMaster, J.; Lubitz, W. Electronic structure of a binuclear nickel complex of relevance to [nife] hydrogenase. *Inorg. Chem.* **2008**, *47*, 11688–11697. [CrossRef]
196. Feng, Y.; Chen, C.; Liu, Z.; Fei, B.; Lin, P.; Li, Q.; Sun, S.; Du, S. Application of a ni mercaptopyrimidine mof as highly efficient catalyst for sunlight-driven hydrogen generation. *J. Mater. Chem. A* **2015**, *3*, 7163–7169. [CrossRef]
197. An, Y.; Liu, Y.; An, P.; Dong, J.; Xu, B.; Dai, Y.; Qin, X.; Zhang, X.; Whangbo, M.H.; Huang, B. Ni(ii) coordination to an al-based metal-organic framework made from 2-aminoterephthalate for photocatalytic overall water splitting. *Angew. Chem.* **2017**, *56*, 3036–3040. [CrossRef]
198. Zhou, T.; Du, Y.; Borgna, A.; Hong, J.; Wang, Y.; Han, J.; Zhang, W.; Xu, R. Post-synthesis modification of a metal–organic framework to construct a bifunctional photocatalyst for hydrogen production. *Energy Environ. Sci.* **2013**, *6*, 3229. [CrossRef]
199. Bloch, E.D.; Britt, D.; Lee, C.; Doonan, C.J.; Uribe-Romo, F.J.; Furukawa, H.; Long, J.R.; Yaghi, O.M. Metal insertion in a microporous metal-organic framework lined with 2,2′-bipyridine. *J. Am. Chem. Soc.* **2010**, *132*, 14382–14384. [CrossRef]
200. Toyao, T.; Saito, M.; Dohshi, S.; Mochizuki, K.; Iwata, M.; Higashimura, H.; Horiuchi, Y.; Matsuoka, M. Development of a ru complex-incorporated mof photocatalyst for hydrogen production under visible-light irradiation. *Chem. Commun. (Camb. Engl.)* **2014**, *50*, 6779–6781. [CrossRef]
201. Dhakshinamoorthy, A.; Asiri, A.M.; Garcia, H. 2d metal–organic frameworks as multifunctional materials in heterogeneous catalysis and electro/photocatalysis. *Adv. Mater.* **2019**, *31*, 1900617. [CrossRef]
202. Fajrina, N.; Tahir, M. A critical review in strategies to improve photocatalytic water splitting towards hydrogen production. *Int. J. Hydrogen Energy* **2019**, *44*, 540–577. [CrossRef]
203. Wang, Z.; Li, C.; Domen, K. Recent developments in heterogeneous photocatalysts for solar-driven overall water splitting. *Chem. Soc. Rev.* **2019**, *48*, 2109–2125. [CrossRef]

MDPI
St. Alban-Anlage 66
4052 Basel
Switzerland
Tel. +41 61 683 77 34
Fax +41 61 302 89 18
www.mdpi.com

Catalysts Editorial Office
E-mail: catalysts@mdpi.com
www.mdpi.com/journal/catalysts

www.ingramcontent.com/pod-product-compliance
Lightning Source LLC
Chambersburg PA
CBHW041955260926
51850CB00019B/2907
* 9 7 8 3 0 3 6 5 1 7 4 8 3 *